Transport
Beyond Oil

Policy Choices for a Multimodal Future

Transport
Beyond Oil

Policy Choices for a Multimodal Future

Edited by John L. Renne and Billy Fields

ISLANDPRESS

Washington | Covelo | London

Copyright © 2013 Island Press

All rights reserved under International and Pan-American Copyright Conventions. No part of this book
may be reproduced in any form or by any means without permission in writing from the publisher:
Island Press, 2000 M Street NW, Suite 650, Washington, DC 20036

Island Press is a trademark of The Center for Resource Economics.

Library of Congress Cataloging-in-Publication Data

Transport beyond oil : policy choices for a multimodal future / edited by John L. Renne and Billy Fields.
    p.   cm.
    Includes bibliographical references and index.
    ISBN 978–1–61091-041–5 (alk. paper) — ISBN 1–61091-041–9 (alk. paper) — ISBN 978–1–61091-043–9
(pbk. : alk. paper) — ISBN 1–61091-043–5 (pbk. : alk. paper) 1. Transportation and state—United States.
2. Sustainable urban development—United States. 3. Transportation—Energy consumption—United
States. 4. Transportation—Energy conservation—United States. 5. Petroleum as fuel—United States.
I. Renne, John L. II. Fields, Billy (Billy M.)
    HE206.2.T6765 2013

Printed on recycled, acid-free paper

Manufactured in the United States of America

10 9 8 7 6 5 4 3 2 1

Keywords: bicycling, biodiesel, car dependence, carbon emissions, climate change, Deepwater Horizon,
freight transportation, gas tax, greenhouse gas emissions, high-speed rail, induced travel demand,
land use planning, mass transit, nonmotorized transportation, peak oil, pedestrian infrastructure,
petroleum consumption, public transportation, transit-oriented development, transportation planning,
transportation policy, urban design, vehicle miles traveled

The editors wish to dedicate this book to all of the people and communities along the Gulf Coast who were and still are impacted by the *Deepwater Horizon* disaster.

Contents

Foreword

Where Have We Come From? Where Are We Going? Interstate 2.0

Like President Obama, a growing number of American people envision an upgraded, higher-speed rail (HSR) transportation system for intercity passengers in the United States. It is a logical and necessary next step forward from President Eisenhower's Interstate Highway System of the 1950s—but proponents have long had a hard time being heard until recently. While HSR has become ultra-politicized, with some governors canceling good programs (Phase 1–funded at $10.1 billion in recent years) and sending money back to the federal government, high-speed rail is an important step in this century's most important transportation infrastructure program.

Many of us remember when the Arab oil embargo took place in October 1973, creating our first energy crisis. Long waiting lines formed at service stations and many stations turned off their lights on the Interstate. They were out of gas! Americans woke up and realized that we had built a mobility system on a finite fossil fuel. I remember that by 1974 people were abandoning their 4,000-pound, eight-cylinder, six-MPG Buicks and lining up to buy the Volkswagen Rabbit diesel. We started to "think small," and solar and wind energies were being discussed. But by the late 1970s, we were seemingly discovering oil under every polar bear in the Arctic. The price of a barrel of oil went from $35 back down to $9–12 a barrel, and by the mid–1980s we were once again well on our way to preferring gas-guzzling muscle cars, SUVs, 400 HP V8s, and $70,000 trucks! Fat City was the way to go—until 2008. Furthermore, research shows the United States had an unwritten transportation policy that declared Americans want "cheap fossil fuel." Any political figure who even talked about raising the gas tax was doomed!

So where are we today with our twenty-first-century global economy? The truly big energy crisis has occurred. Oil rose to $140-plus per barrel. Gasoline/diesel went to $5 per gallon. Since 2010, oil prices per barrel have remained $80–100 per barrel, which has clobbered the transportation industry, especially aviation. In recent years our Big Three car manufacturers were shattered, and our economy is still on life support. Congress cannot keep prices low by legislation. Global economic chaos will result if just one major oil-producing nation has some sort of calamity.

We can no longer afford the lavishness of the past. As soon as possible, this nation has got to radically change the way People and Freight move in order to avoid long-term economic decline. One need only look at our demographics and our growing population density. When I was thirty, there were 130 million people in the United States. By 2040, there will be 400 million people in the United States and North America will have a population of well over half a billion people! As we finish the first decades of this new century, the old order of "business as usual" is not working. What is the biggest public-works project that can ensure prosperity in this century? Last century it was building Interstate 1.0—the Interstate Highway System—with its 43,000 miles of grade-separated, four-lane highways. It served millions of cars and trucks, and it fed into thousands of small, busy airports with commuter airplanes as well as huge hub airports with large passenger planes going long distances to big cities. In the 1970s and 1980s the airlines expanded, in part because jet fuel prices were about 40–60¢ per gallon, with no tax! Western man built a huge, gaudy, wasteful, polluting transportation system on this cheap oil; it employed millions of people and we all prospered. But that is all over now, and the good old days of cheap fuel will not return.

Where Are We Going?
So what do we do now? What major public-works project can we implement in this century that will help keep our 400 million people working, will promote a prosperous economy, and will build a long-lasting, sustainable transportation system? My answer is that we build "Interstate 2.0." I initially said it should be 20,000 miles of grade-separated, higher-speed rail. It really should be 30,000 miles and should use the huge, wide, existing—and paid for—rail rights-of-way in partnership with the private freight railroads and the states. We should give the private railroads their 25 percent investment tax credit to encourage them to upgrade and double- and triple-track their main lines in order to increase speeds and double freight/passenger capacity. States could build or lease high-speed track on their ROWs to run new, modern, intermodal passenger trains. Most of these high-speed tracks should be grade-separated as were the Interstate highways. Our objective is to enable Amtrak and its partners to run

frequent and safe 110–125 MPH passenger trains. We have the technology to do this with a high degree of safety. It will cut the number of highway fatalities and drastically reduce the wear and tear on the highways as well as the cost of maintaining them.

Intermodal and high-speed passenger-rail visionaries have finally been heard by the president. A huge 2.0 work program puts America on the way to creating an "ethical" intermodal freight and passenger transportation network. We can electrify it by midcentury. It will then truly be an "ethical and sustainable" system. President Obama will be the twenty-first century's Eisenhower because he will have created "Interstate 2.0," a high-speed rail network reconnecting our center cities, major airports, and ports—thus recapturing the vital role of the intercity train, bus, and transit industries.

By the way of explanation: *An "ethical transportation system" is one that (1) does not injure or kill; (2) does not pollute and is environmentally benign; (3) does not waste fuel; and (4) does not cost too much. It uses the strengths of each mode of transportation.* We can build a twenty-first-century intermodal transportation system using the "steel wheel and steel rail" as the fundamental element. Early in this century we can electrify all of North American rail, thereby providing a new source of energy for our transportation system.

We have started. This is Phase 1—$10.1 billion and a number of federally designated, high-speed rail corridors in different regions of the country. Amtrak has survived and will show the American people that a truly integrated, intermodal, passenger transportation system is coming. By using our existing freight-rail ROWs and not destroying more green fields, we can actually create a much better transportation system than Europe.

It is an exciting new era that we are entering. Thank you.

—Gilbert E. Carmichael
Founding Chairman, ITI Board of Directors, Intermodal Transportation Institute, University of Denver
President, Missouth Properties
Federal Railroad Administrator under President George H. W. Bush, 1989–93

Acknowledgments

The editors wish to thank each of the contributors. We are indebted to Heather Boyer, Senior Editor at Island Press, for all of her hard work and guidance. We are also grateful to Courtney Lix at Island Press and Tara Tolford at the University of New Orleans for helping us get to the finish line. Finally, we wish to thank our families, including Kara, Tara, and Keagan, for their patience and the forgone family time due to our work on this book.

Moving from Disaster to Opportunity

Transitioning the Transportation Sector from Oil Dependence

John L. Renne and Billy Fields

In the spring and summer of 2010, America was transfixed by the image of oil spewing into the Gulf of Mexico from the collapsed remnants of the *Deepwater Horizon* oil platform. Images of majestic pelicans floundering in oil and the personal stories of the eleven crew members who lost their lives when the oil platform exploded were interspersed with camera shots of the seemingly never-ending stream of oil emanating from a broken pipe a mile below the Gulf's surface. While the *Deepwater Horizon* disaster became the poster child for corporate greed and neglect, few considered how America's transportation dependence on oil helped stimulate demand for the oil pouring into the Gulf. Seventy percent of all oil consumed in the United States goes to the transportation sector, mostly powering single-occupant vehicles that Americans use for 82 percent of all trips.[1]

To put this in the context of the *Deepwater Horizon* disaster, imagine that 70 percent of the 68,000 square miles of oil that was floating in the Gulf of Mexico was destined to be consumed by America's transportation sector.[2] The area covered by the oil intended for the transportation sector would cover an area slightly larger than the entire state of Pennsylvania (47,600 square miles). Perhaps more shocking is that, despite the massive amount of oil spilled in the Gulf of Mexico, the quantity used by the transportation sector alone would be consumed in just under three days.[3]

The shocking images from 2010 (e.g., figs. 0.1 and 0.2) have now mostly given way to a slow-motion aftermath of impacts. The Official Selection Documentary of the 2011 Cannes Film Festival, *The Big Fix*, details how pervasive the disaster was—and still

Figure 0.1   Oil-spill impacts on the white sandy beaches of Gulf Shores, Alabama. (Source: istockphoto.com.)

Figure 0.2   Oil-spill impacts on wildlife. Louisiana's coast is a critical stopover habitat for hundreds of species of nesting and migratory seabirds and other waterfowl, including many of North America's most at-risk species. (Source: US Coast Guard, www.ecy.wa.gov /programs/spills/Special_Focus /BP_LA_Oilspill/photo_gallery /wildliferescue_pelican.jpg.)

is—on local economies trying to recover, and it illuminates the uncertain long-term environmental and health impacts on marine populations and coastal communities.[4]

With the disaster receding into our collective memories, proposals for transformative policy response have now given way to inaction. The policy window for change from this disaster has closed, while the demand for oil for the transportation sector continues unabated (see fig. 0.3). The impacts of oil extraction and dependence are often perceived as necessary evils that must be accepted in order to maintain modern standards of living. We are repulsed by the string of oil disasters, but feel powerless to find transformative solutions that can decrease oil dependence.

The chapters in *Transport Beyond Oil* were crafted to provide a data-driven platform to discuss realistic opportunities to transition the transportation sector away from oil dependence. The book addresses the systemic problems underlying America's oil dependence and provides detailed policy alternatives that can help to chart a new course. The chapters throughout the book show how the United States can alter its course of transportation oil dependence and move toward a future with a new economic foundation, greater livability, and an improved environment for the twenty-first century.

Peak Oil and Extreme Oil Impact on Household Budgets and Environmental Disasters

Petroleum is both a scarce natural resource and a ubiquitous product available at your neighborhood gas station for a unit cost cheaper than that of the bottled water you can also purchase there. The concept of peak oil helps to explain how scarcity and availability interact. Proponents of the peak oil concept argue that the Earth has about 50 percent of its oil reserves remaining; however, production capacity has peaked and supplies will not be able to match even current demand. Anyone who has taken a basic economics course could point out that a market with high demand and limited supply will result in upward pressure on the cost of a commodity. Since oil is used universally to manufacture and transport virtually all other commodities, an increase in oil cost results in an increase in the production cost of all other goods. While the peak oil concept was once hotly debated, scientists, government officials, and oil companies have generally accepted peak oil as a fact, along with the implication of increasing prices. The question of when the global peak will occur is still subject to some debate, but most agree that, despite recent discoveries, the global peak will occur during the first quarter of the twenty-first century.

Related to peak oil is the concept of extreme oil. The extraction of the first half of the world's oil supplies, which occurred mainly during the twentieth century, was relatively easy as this oil was located in large oil reserves. The second half of the world's reserves is much harder and more expensive to extract. Therefore, higher extraction costs will ultimately result in higher consumer prices, adding to the price of petroleum-based transport. And as the *Deepwater Horizon* disaster made clear, the search for extreme oil through more complex extraction technologies also results in increased potential for accidents that are difficult to control. Extracting oil from a mile below the ocean's surface can result in mega-disasters that could take months to control, with impacts possibly lingering for decades. Meanwhile, demand continues to grow.

Independently, peak oil and extreme oil could have a significant inflationary impact on the cost of petroleum-based products for Americans. The combined effect could result in serious national economic instability. The average American spends 18 percent of their household budget on transportation. An additional 12 percent is spent on food, 9 percent on home heating and household operations, and 10 percent on apparel, personal care products, and home furnishings. Thus, approximately half of all purchases are directly related to oil prices, since petroleum is a significant raw material for the manufacture and transport of all of these products.[5]

While financial demands of transportation oil dependence will strain individual and community budgets, the environmental impacts of oil dependence provide both significant short- and long-term challenges. As noted above, the *Deepwater Horizon* Oil Spill in 2010, which served as the impetus for writing this book, was one of the largest global environmental disasters in history. However, the environmental consequences of our automobile-dependent transportation system extend well beyond oil spills. Climate change resulting from greenhouse-gas emissions, air pollution, and water pollution are all serious threats to our local and global ecosystems. These impacts are addressed in greater detail in subsequent chapters.

How Does America Address Oil Addiction and Car Dependence?

As America seeks to emerge from the Great Recession, our national and state highway trust funds are nearing bankruptcy. With hyper-partisanship dominating Congress, leaders have been unable to agree on a plan for maintaining the national transportation system beyond status quo policies. Economic stagnation, political gridlock, and the status quo result from decision makers trying to patch a broken system. As Richard Florida writes,

> our political and business leaders have utterly failed to appreciate and engage this economic transformation. They continue to look backward, with futile attempts

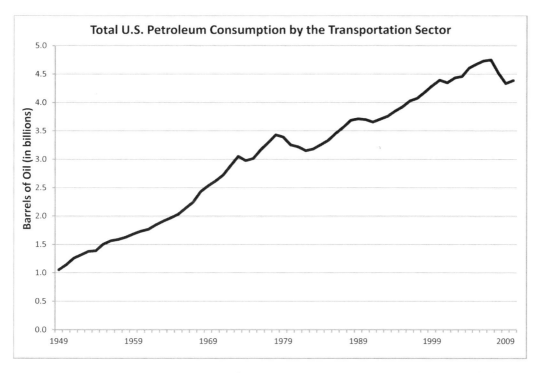

Figure 0.3   Transportation-sector petroleum-consumption estimates, 1949–2010 (in billion barrels of oil). (Source: US Energy Information Administration, Annual Energy Review, October 2011.)

to resuscitate the dysfunctional system of banks, sprawl, and the inefficient and energy-wasting way of life that was the underlying cause of the crisis.[6]

Some scholars contend that we are entering a period where new transportation innovations are needed to break our oil addiction.[7] As in the period after the Long Depression of 1873 and the Great Depression of the 1930s, the economy and society are presently experiencing fundamental changes that can enable new growth and productivity to emerge. These fundamental changes are putting pressure on our transportation sector. Yet change can be slow and painful, especially when we consider how addicted America is to the automobile and its main source of power—oil. Our ability to solve these national problems will determine either the recovery of America's economy or the devolution of our status as a superpower.

The oil consumption paradox exemplifies how problematic our addiction is. Our nation funds transportation through a highway trust fund with tax dollars received from gasoline sales. The less we drive, the more broke the trust fund becomes. So should our politicians encourage us to drive more, by investing in wider and more expansive highways, which in turn will induce more vehicle-miles traveled? Or should we seriously consider policies that encourage more walking, bicycling, and mass-transit use across cities, which will reduce driving? Should we seriously consider electrically operated buses, trains, and automobiles that decrease our need for oil? Policies that reduce our oil consumption diminish our nation's ability to afford to invest in our transportation system. This is the first conundrum that must be solved for our transport system to move beyond oil as we seek to adapt our economy for new growth.

Recent Scholarship Related to Transportation and the Oil Crisis

There are vast amounts of literature on the consequences of our oil addiction, including a number of recent sources that address transportation and policy. (The references in this book contain numerous such sources.) A few that we would like to highlight, because they focus on the nexus of the impending oil crisis and the future for cities and the transportation system, are *The Long Emergency: Surviving the Converging Catastrophes of the Twenty-First Century* (2005),[8] *Lives per Gallon: The True Cost of Our Oil Addiction* (2006),[9] *Resilient Cities: Responding to Peak Oil and Climate Change* (2009),[10] *Two Billion Cars: Driving Toward Sustainability* (2009),[11] and *Transport Revolutions: Moving People and Freight Without Oil* (2010).[12]

Transport Beyond Oil: Policy Choices for a Multimodal Future seeks to add to this recent scholarship and provide detailed, policy-relevant pathways to begin the transition to a world beyond oil.

Overview of Transport Beyond Oil

Chapters in *Transport Beyond Oil* are presented in three sections:

    Part 1: Petroleum Consumption Impacts and Trends
    Part 2: Transportation and Oil Dependence: A Modal Analysis
    Part 3: Moving Forward

In part 1, Debbie Gordon and David Burwell's chapter, "The Role of Transportation in Climate Disruption," discusses the sobering relationship between transportation and climate change. In "Oil Vulnerability in the American City," Neil Sipe and Jago Dodson describe how households are vulnerable to price fluctuations stemming from oil prices. Next, Todd Litman focuses his chapter on the "Full Cost Analysis of Petroleum Consumption."

Part 1 continues with Robert Noland and Christopher Hanson discussing "How

Does Induced Travel Affect Sustainable Transportation Policy?" This is followed by Deron Lovaas and Joanne Potter's "Bending the Curve: How Reshaping US Transportation Can Influence Carbon Demand," which provides ideas for a suite of policy options that could result in a brighter future that is less oil dependent.

In part 2, Bradley Lane focuses on transit in his chapter "Public Transportation as a Solution to Oil Dependence." Projjal Dutta also examines transit in "Taking the Car Out of Carbon: Mass Transit and Emissions Avoidance." Petra Todorovich and Edward Burgess examine "High-Speed Rail and Reducing Oil Dependence." Simon McDonnell and Jie (Jane) Lin discuss "The Challenges and Benefits of Using Biodiesel in Freight Railways." Kevin Mills examines "Healthy, Oil-Free Transportation: The Role of Walking and Bicycling in Reducing Oil Dependence," and Alan Drake's chapter focuses on "Building an Optimized Freight Transportation System."

Part 3, the final section of the book, seeks to map a path forward. Peter Newman writes a chapter on "Imagining a Future Without Oil for Car-Dependent Cities and Regions." John Renne discusses "The Pent-Up Demand for Transit-Oriented Development and Its Role in Reducing Oil Dependence." Jeff Kenworthy examines whether our cities are "Deteriorating or Improving? Transport Sustainability Trends in Global Metropolitan Areas." Billy Fields and Tony Hull focus on "Policy Implications of the Nonmotorized Transportation Pilot Program: Redefining the Transportation Solution." Finally, Billy Fields, John Renne, and Kevin Mills propose "From Potential to Practice: Building a National Policy Framework for Transportation Oil Reduction."

Henry Ford's words—"Don't find fault, find a remedy"—ring true today for our transportation industry. Perhaps if America's greatest entrepreneur were alive today, he would find a way to once again revolutionize our transportation system and solve the large challenges we face while opening a new market for the next generation.

Fortunately, the contributors of *Transport Beyond Oil* are able not only to articulate one of the most important challenges facing our society but also provide a remedy for a sustainable, multimodal future.

Notes

1. Federal Highway Administration, *National Household Travel Survey* (Washington, DC: US Department of Transportation, 2009).

2. Justin Gillis, "An Oil Slick to Rival Oklahoma," *New York Times*, July 28, 2010.

3. Joel Achenbach and David A. Fahrenthold, "Oil Spill Dumped 4.9 Million Barrels into Gulf of Mexico, Latest Measure Shows," *Washington Post*, August 3, 2010; US Energy Information Administration, "Petroleum Statistics" (Washington, DC: US Department of Energy, 2011), www.eia.gov/energyexplained/index.cfm?page=oil_home#tab2.

4. See: www.imdb.com/title/tt1939753/.

5. US. Bureau of Labor Statistics, "Consumer Expenditures" (Washington, DC: US Department of Labor, April 2009), www.visualeconomics.com/how-the-average-us-consumer-spends-their-paycheck/.

6. Richard Florida, *The Great Reset* (New York: Harper, 2011), xi.

7. Richard Gilbert and Anthony Perl, *Transport Revolutions: Moving People and Freight Without Oil* (Gabriola Island, BC: New Society Publishers, 2008); Peter Newman, Timothy Beatley, and Heather Boyer, *Resilient Cities: Responding to Peak Oil and Climate Change* (Washington, DC: Island Press, 2009); Florida, *The Great Reset*.

8. James Howard Kunstler, *The Long Emergency: Surviving the Converging Catastrophes of the Twenty-First Century* (New York: Atlantic Monthly Press, 2005).

9. Terry Tamminen, *Lives per Gallon: The True Cost of Our Oil Addiction* (Washington, DC: Island Press, 2006).

10. Newman et al., *Resilient Cities*.

11. Daniel Sperling and Deborah Gordon, *Two Billion Cars: Driving Toward Sustainability* (Oxford, UK: Oxford University Press, 2009).

12. Gilbert and Perl, *Transport Revolutions*.

Part 1

Petroleum Consumption Impacts and Trends

1 The Role of Transportation in Climate Disruption[1]

Deborah Gordon and David Burwell

The Earth's rapidly warming temperatures over the past several decades cannot be explained by natural processes alone. The science is conclusive: both man-made and natural factors contribute to climate change. Human activities—fossil-fuel combustion in transportation and other sectors, urbanization, and deforestation—are increasing the amount of heat-trapping gases in the atmosphere. These record levels of greenhouse gases are shifting the Earth's climate equilibrium.

Climate impacts differ by sector. On-road transportation—cars and trucks—has the greatest negative effect on climate, particularly in the short term.[2] This is primarily because of two factors unique to on-road cars and trucks: (1) nearly exclusive use of petroleum fuels, the combustion of which results in high levels of the principal climate-warming gases (carbon dioxide, ozone, and black carbon); and (2) minimal emissions of sulfates, aerosols, and organic carbon from on-road transportation sources to counterbalance warming with short-term cooling effects.

Despite its leading role as the largest source of short-term climate forcing, transportation is not shouldering its responsibility in reducing greenhouse-gas emissions.[3] Moreover, the US (and global) transportation situation is especially problematic, given the dependence on oil that characterizes this sector today. There are too few immediate mobility and fuel options in the United States beyond oil-fueled cars and trucks. Moreover, many of the new oils being tapped—oil sands and shale oil, for example—emit more carbon than conventional oil. Clearly this sector, as a major contributor to climate change, should be the focus of new policies to mitigate warming. Government must lead this effort, as the market alone cannot bring about the transition away from cars and oil.

Policy makers need to remember four essential findings when developing new strategies for ensuring that the United States maintains its Copenhagen commitment to reduce greenhouse-gas emissions (17–20 percent below 2005 levels by 2020) while also retaining its leadership position in the global economy. First, on-road (car and truck) transportation is an immediate high-priority target in the short term for reducing greenhouse-gas emissions and mitigating climate change in the United States and around the globe. Second, the transportation sector is responsible for high levels of long-lived carbon dioxide and ozone precursor emissions that will warm the climate for generations to come. Third, the United States (and other nations) must transition quickly to near-zero greenhouse-gas-emission (GHG) cars and trucks, largely through low-carbon electrification for plug-in vehicles. And finally, America's transportation culture must adapt to relying less on fossil fuels through technological innovation, rational pricing, and sound investments that expand low-carbon mobility choices and that fundamentally shift travel behavior.

Climate is a condition that will define the twenty-first century, especially global mobility. There are reasons to be optimistic about the challenges ahead. Climate scientists find that cutting on-road transportation climate-changing and air-pollutant emissions would reduce climate forcing and benefit public health in the near term. Supporting a new, low-carbon, location-efficient, productive, and high-growth economy will be key to thriving in an increasingly competitive global marketplace.

Climate as a Condition

According to the National Oceanic and Atmospheric Administration (NOAA) and the National Aeronautics and Space Administration (NASA), global surface temperatures have risen by 0.6°C since the middle of the twentieth century. The current decade has been the warmest worldwide on record, 0.2°C warmer than the 1990s.[4] According to the US Environmental Protection Agency (EPA), the evidence of the Earth's warming is clear.[5]

The Earth's global average temperature is projected to rise 1.7–3.9°C by 2100, and continue to warm in the twenty-second century.[6] Scientists are certain that human activities are changing the composition of the atmosphere and that increasing the concentration of greenhouse gases will change the planet's climate. But they are still working to better understand the precise mechanisms of climate change, how much or at what rate temperature will increase, and what the likely effects will be.

Still, scientists warn that the floods, fires, melting permafrost and ice caps, torrid heat, droughts, tornadoes, and other forms of extreme weather witnessed in the past couple of years are signs of troubling climate change already under way.[7] As shown in figure 1.1, about two new high-temperature records were set for every one low-temperature record during the 2000s. And the ratio of record high to record low temperatures has increased since the 1960s. Scientific evidence strongly suggests that man-made increases in greenhouse gases account for most of the Earth's warming over the past fifty years.

The National Research Council reports in *Climate Stabilization Targets: Emissions, Concentrations, and Impacts of Decades to Millennia* that carbon dioxide (CO_2) accounts for more than half of the current effect on the Earth's climate. Scientists are more concerned about the climate effects of anthropogenic (man-made) carbon dioxide emissions than any other greenhouse gas.[8] The atmospheric concentration of carbon dioxide is at its highest level in at least 800,000 years.[9]

Carbon dioxide flows into and out of the ocean and biosphere. Man-made carbon dioxide creates net changes in these natural flows, which accumulate over time; such extreme persistence is unique to carbon dioxide among major warming gases. Black carbon and greenhouse gases, such as methane, can also affect the climate, but these

Figure 1.1   Ratio of US record high to low temperatures. (Source: National Center for Atmospheric Research, November 12, 2009, www2.ucar.edu/news/record-high-temperatures -far-outpace-record-lows-across-us.)

changes are short-lived and are expected to have little effect on global warming over centuries or millennia.

But even if carbon dioxide emissions were to end today, scientists expect that changes to Earth's climate that stem from carbon dioxide will persist and be nearly irreversible for thousands of years. Scientists' best estimate is that for every 1,000 gigatonnes (GtC) of anthropogenic carbon emissions, average global temperatures will increase 1.75°C.[10] Therefore, each additional ton of carbon dioxide released into the atmosphere forces warming.

Untangling the Connection Between US Transportation and Climate

Direct Greenhouse Gas Emissions from Transportation

The direct greenhouse-gas (GHG) emissions—carbon dioxide, methane, nitrous oxide, and synthetic halocarbons—can be accounted for in different ways. Carbon dioxide, which has an atmospheric lifetime of at least 100 years, dominates direct GHG emissions from energy-related activities, primarily due to fossil fuel combustion. Regardless of the method chosen, the direct GHG emissions are measured in terms of their carbon dioxide-equivalent ($CO_{2\,Eq}$.) levels based on their relative ability to force climate warming.

Climate researchers suggest that climate science needs to shift from looking at the impact of individual chemicals to examining output by economic sector.[11] Each economic sector emits a unique portfolio of gases and aerosols that affect the climate in different ways over different time frames. The IPCC disaggregates emissions into the self-defined sectors, including energy, industrial processes, solvent and other product use, agriculture, land use,[12] and waste. When the energy sector is further disaggregated and fuel-combustion related emissions are accounted for, the following economic sectors are considered: transportation, industry, commercial, and residential. Transportation edges out industry as the largest source of carbon dioxide emissions and thus as a key driver of climate change.[13]

Each sector's share of direct GHG emissions can be reported with or without electric power generation included. Electric power supplies energy to most of the economic sectors, except for transportation. When broken out, the electric power industry generates more direct carbon-equivalent climate gases overall than any economic sector. Transportation has the second-highest direct GHG emissions, followed by industry, commercial, and residential sectors. The agriculture sector is reported in GHG inventories but is not included here, given its large emission "sinks" that counteract emission sources.

Another way to evaluate direct climate-gas-emission inventories is to distribute electricity-related emissions based on actual use by each economic sector. The transportation sector uses essentially no electricity (actually 0.003 percent) but still has nearly the same direct GHG emissions as industry (about 30 percent).[14]

Air Pollutant Climate Precursor Emissions from Transportation

In addition to carbon dioxide and the other direct GHGs mentioned above, the transportation sector accounts for a significant portion of additional emissions that react to form air pollution (known as precursors).[15] These emissions, detailed below, affect the climate through a variety of complex chemical reactions.[16] The transportation sector is responsible for the majority of air pollutant precursor emissions—a full 85 percent of carbon monoxide (CO), 50 percent of black carbon (BC), 34 percent of particulate matter of 2.5 microns ($PM_{2.5}$), 55 percent of nitrogen oxides (NO_x), and 41 percent of non-methane volatile organic compounds (NMVOC). The utility sector, on the other hand, is responsible for 86 percent of total sulfur dioxide (SO_2) emissions.[17] In the United States, on-road transportation is not responsible for SO_2 emissions. These figures represent current emission levels from burning conventional oil as the primary fuel source. The increased use of unconventional oil (oil sands, oil shale, ultra-heavy and ultra-deep oils, and coal-to-liquids) is likely to increase direct GHG emissions and air pollution precursors, resulting in an even greater air pollution emission share from tomorrow's transportation sources.

On-road transportation sources emit both NMVOC and NO_x in large amounts. These ozone precursors react to form ozone, or what the public often calls *smog*. Carbon monoxide is produced when carbon-containing fuels fail to fully combust in cars and trucks, and nitrogen oxides (NO and NO_2) are created from the nitrogen in the air when burning fossil fuels.

Black carbon (BC) is another air pollutant precursor that acts as a climate agent. BC is rarely measured in its pure form. Instead, it is part of particulate matter, which constitutes a broad array of carbonaceous substances, sometimes referred to as *soot*. The incomplete combustion of fuel (from transportation and other sources) results in black carbon (and organic carbon), fine particles that are suspended in the atmosphere.[18] These particles are identified by their size: $PM_{2.5}$ and PM_{10}, or less than 2.5 μm (micrometers) and 10 μm, respectively. Diesel fuel combustion, moving freight in heavy-duty trucks, is the major source of black carbon in the United States, but the EPA does not report black carbon (PM) emissions in its GHG Trends Reports. However, PM is inventoried for air pollution modeling.[19]

Air pollutants and direct greenhouse-gas emissions are intimately connected atmospherically. While regional air pollutants influence climate change, a warmer climate can also exacerbate air pollution. This effect occurs because heat accelerates many air-pollutant reactions.[20] Thus, an increase in global warming precipitates an increase in regional pollution, and vice versa.

Probing the Relationships Between US Transportation and Energy

US Transportation, Energy, and Oil Use

Energy consumption and climate change are inextricably linked; the energy sector in its entirety accounts for 86 percent of total direct GHG emissions.[21] The energy requirements of each economic sector (transportation, industry, commercial, and residential) are responsible for the bulk of all man-made climate-change gases. Transportation represents a significant portion of emissions in the Intergovernmental Panel on Climate Change (IPCC) energy sector.

In 2010, the transportation sector consumed 27.5 quads (quadrillion, or 10^{15} BTU) of direct energy, mostly in the form of refined liquid fuels, chiefly gasoline and diesel fuel. Transportation's share of energy consumption is similar to its share of greenhouse-gas emissions, at 34 percent. The linkages between energy use and climate gases are evident in all economic sectors.

Unlike other economic sectors, transportation runs nearly exclusively on petroleum, which fuels 94 percent of this sector's energy demands.[22] In 2011, US. mobility (cars, trucks, airplanes) required nearly 14 million barrels per day (mbpd) of oil, out of 18.8 mbpd total US oil consumption. Transportation used three times more oil than did all industries combined. And the transportation sector consumed ten times more oil than the commercial, residential, or utility sectors.

Energy, Oil Use, and Carbon Emissions

There is near parity between energy use and emission of the principal greenhouse gas, carbon dioxide. Essentially all of the carbon contained in fossil fuels, which are hydrocarbons, is converted to carbon dioxide when burned. Solar, wind, hydroelectric, geothermal, wave, and nuclear energy contain no carbon and, therefore, have no direct effect on GHG inventories. Biofuels, on the other hand, contain carbon. Biofuels, a myriad of plant- and waste-based fuels, have a complex relationship to the climate. Their GHG emissions depend on their individual chemistries, how they combust, and even how their feedstocks are grown.

The amount of carbon released into the atmosphere is primarily determined by the fuel's carbon content.[23] Today, the on-road (car and truck) transportation system runs almost exclusively on gasoline and diesel fuels. An average gallon of gasoline, once combusted in air, converts its carbon to 19.4 pounds of CO_2 per gallon of gasoline consumed (8.8 kg/gallon). Diesel, the fuel primarily used in heavy-duty trucks and off-road vehicles, has 22.2 pounds CO_2 per gallon (10.1 kg/gallon).[24]

Conventional crude-oil-derived fuels are beginning to be replaced by new *unconventional* oils, such as bitumen (oil sands), tight shale oil, kerogen (oil shale), and coal-to-liquids. Unconventional oils contain as much as triple the carbon as today's crude oil.[25] Moreover, new oils require more energy for their extraction and processing.

Given new policy mechanisms in the longer term, non-oil fuels could replace new oils. There are more than 100 fuel-production pathways and over 70 vehicle and fuel-system pairings, each with its own climate emission impact, as illustrated in fig. 1.2. Again, the carbonization of future fuels will vary, depending on the fuel source and production pathway chosen. Electricity generated by burning coal produces high carbon emissions, but electricity from many renewable and nuclear sources has zero emissions. A comprehensive fuel-cycle analysis offers the best comparison of total emissions. But this calculation must consider *all* of the carbon in the base fuel resource, whether it yields high-value transport fuels (gasoline, diesel, or jet fuel) or low-value industrial fuels (petroleum coke or residual oils).

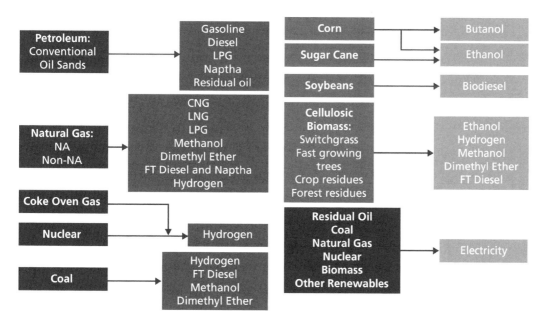

Figure 1.2   Sampling of transportation fuel-production pathways. (Source: M. Wang, "Well-to-Wheels Greenhouse Gas Emissions of Alternative Fuels" [Argonne National Laboratory, 2007], www.nga.org/files/pdf/0712alternativefuelswang.pdf.)

Considering Fuel-Cycle Emissions from Various Forms of Transportation Energy
Fuel-cycle emissions consider all parts of the transportation energy process that can produce greenhouse-gas emissions. Also termed "well-to-wheel" emissions, fuel-cycle emissions start at the wellhead, where fuel is extracted, and end at the tailpipe, where emissions emerge after fuel is combusted in an engine, as illustrated in figure 1.3.

The first part of the process, "well-to-tank," includes fuel extraction, initial fuel processing, intermediate fuel transport, finished fuel production and refining, and distribution and marketing. Reducing total *upstream* GHG emissions are the responsibility of oil companies whose fuel-stock selections, operations, processing, and distribution networks determine carbon intensity. Greater accuracy, however, is needed to depict the carbon intensity of new oil inputs economy-wide, accounting in the aggregate for all of their by-products.

As oils transform, so too will their emissions. Attributing carbon flows from each of the upstream processes and by-products (such as petroleum coke, a coal-like product removed in the process of upgrading oil sands) to downstream petroleum products (such as gasoline) is a huge challenge in terms of accurate carbon accounting. The knowledge gaps on unconventional oils are extensive. New economy-wide methodologies need to be developed to measure the carbon emissions associated with a growing number of unconventional oil processes and products.[26]

The second part of the fuel-cycle process, "tank-to-wheels," encompasses all emissions during vehicle operation and refueling. These GHG emissions are determined

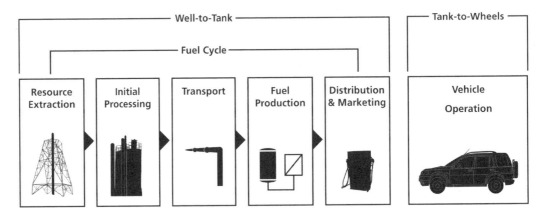

Figure 1.3   Transportation fuel cycle. (Source: Michael Wang et al., "Well-to-Wheels Analysis of Advanced Fuel/Vehicle Systems—A North American Study of Energy Use, Greenhouse Gas Emissions, and Criteria Pollutant Emissions" [US DOT, May 2005], www.transportation.anl.gov/pdfs/TA/339.pdf.)

by automakers in terms of vehicle fuel economy and also by motorists in determining how to efficiently operate and maintain their vehicles.

Studies of fuel-cycle emissions from on-road transportation have found that emissions vary greatly depending on their fuel pathway.[27] Electric vehicles charged with renewable or nuclear fuels can result in near-zero emissions. Plug-in hybrid vehicles can reduce emissions by about 45 percent if powered by electricity generated with limited coal. Carbon dioxide emissions are recycled through plant photosynthesis, so biofuels can provide large reductions (more than 60 percent compared with gasoline), depending on fuel source and processing intensity. Natural gas also reduces GHG emissions, although its potential reduction is much greater when used to generate power for electric vehicles (more than 70 percent) compared with burning compressed natural gas (CNG) directly in vehicles (less than 30 percent).

However, when oil sands made from bitumen are used to power vehicles, fuel-cycle carbon emissions increase substantially compared to conventional oil. One-half of each barrel of bitumen can contain high carbon residual fuel oil, yet these co-products are not uniformly included in fuel-cycle analyses to date. There are numerous uncertainties to be addressed, and further analysis is required to determine just how much more carbon is produced in the full fuel-cycle for unconventional oils.[28]

For heavy-duty vehicles, electric motors provide the most significant benefits (approximately 50 percent), followed by hydrogen fuel cells and CNG. Low-carbon biodiesel can yield a 10–20 percent carbon reduction. Unconventional oils and coal fuels—tar sands, coal-to-liquids, and electric vehicles powered by coal-fired utilities—increase fuel-cycle GHG emissions by 75 percent or more.

Air pollutant emissions that contribute to climate change also vary depending on the alternative fuel and the fuel cycle selected. Ethanol production and use can increase NO_x and particulate emissions, and natural-gas-based hydrogen pathways can reduce criteria pollutant emissions. Heavy-duty vehicles with electric drive have lower particulate and NO_x emissions than do those powered by diesel. And renewable electricity used in electric vehicles and plug-in hybrids can reduce air pollutants that serve as indirect GHG emissions.

Scientists stress that changes in agricultural land use have a large-scale impact on the evaluation of biofuel pathways. Even the sustainable agricultural practices that can be used in biofuel production need to be investigated in order to account for actual fuel-cycle GHG emissions. The prevention of tropical deforestation associated with fuel production, for example, must be incorporated into efforts to promote low-carbon alternative-fuel use. In that case, palm oil could be a low-carbon pathway, but only if grown sustainably, without deforestation.[29]

Distinguishing Climate Impacts Between Transportation and Other Sectors

Each economic sector emits a unique portfolio of gases and aerosols that affect the climate in different ways over different time frames. Scientists have recently determined that transportation is a key driver of climate change. Research by NASA's Goddard Institute for Space Studies suggests climate science needs to shift from looking at the impact of individual chemicals to economic sectors.

Until now, climate effects have been investigated with regard to the impacts of individual greenhouse gases. A comprehensive, sector-by-sector approach, however, is actually more revealing. Sector profiles differ greatly when one considers each sector's climate impacts of tropospheric ozone, fine aerosols, aerosol-cloud interactions, methane, and long-lived greenhouse gases.

In a seminal NASA paper, published by the *Proceedings of the National Academy of Sciences*, Nadine Unger and her colleagues described how they used a climate model to estimate the impact of thirteen sectors of the economy on the climate in 2020 and, more long term, in 2100.[30] They based their calculations on real-world inventories of emissions collected by scientists around the world, and they assumed that in the future those emissions would stay relatively constant at their 2000 levels.

Breaking the massive energy sector into its subsectors is the key because each produces a different, complex mixture of direct GHGs and air pollutant precursor emissions. Some cause longer-term warming; others cause shorter-term cooling, as shown in figure 1.4.

Framing climate change by economic activity provides a better understanding of how human activities affect climate and over what time frames. This approach can foster the development of smart climate policies that identify new opportunities for controlling man-made warming.

Burning fossil fuels with high sulfur contents—particularly coal, high-sulfur diesel, and fuel oil—releases sulfates, which cause short-term cooling by blocking radiation from the sun and making clouds brighter and longer-lived.[31] In the short term, the cooling from sulfates outweighs the warming from carbon dioxide, so the net impact of heavy industries and coal-fired power production cools the climate. Still, carbon-dioxide emissions from coal-fired power generation are so massive that even their long-term warming effects greatly outweigh the short-term sulfate cooling effects on time scales relevant for climate change.[32]

Just because an activity causes cooling in the short term does not mean it is harmless or "good" for the climate. Increasing emissions from coal-fired power generation is detrimental in the long term for the climate and air quality. For example, sulfate aerosols from power plants pose serious air-quality problems, including acid rain, and

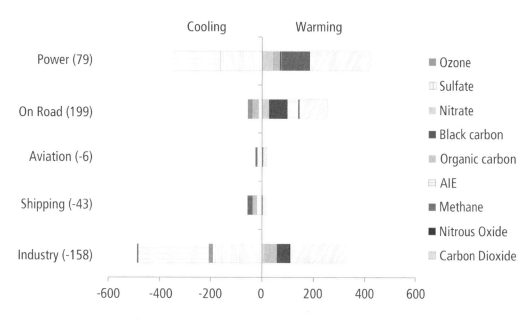

Figure 1.4 Radiative forcing due to perpetual 2000 global emissions, grouped by sector (in 2020). Emission trajectories in this graph assume business-as-usual (year 2000) transport activities powered by oil, without significant EV replacement. (Source: N. Unger, T. C. Bond, J. S. Wang, D. M. Koch, S. Menon, D. T. Shindell, and S. Bauer, "Attribution of Climate Forcing to Economic Sectors," *Proc. Natl. Acad. Sci.* 107 [2010]: 3382–87, doi: 10.1073 /pnas.0906548107.)

affect regional climate in other detrimental ways, such as changing the Earth's water cycle.

As the number of coal-fired power plants in the United States and worldwide increases, the power sector will overtake transportation as the leading climate disruptor, as indicated in figure 1.5. Therefore, cutting coal-fired power-plant (and other industrial) emissions will be crucial in mitigating climate change in the longer term.[33]

During the twenty-first century, on-road transportation is expected to be a leading climate-forcing activity worldwide. Cars and trucks emit almost no sulfates (cooling agents) but are major emitters of carbon dioxide, black carbon, and ozone—all of which cause warming and are detrimental to human health. US on-road transportation is responsible for 40 percent of global on-road climate warming ("radiative forcing" in climate terms).

On-road transportation in the United States (and abroad) is a prime emission-reduction opportunity, at least through 2050. Electric power generation and industry are also high priorities, but they will be more problematic within longer time frames.

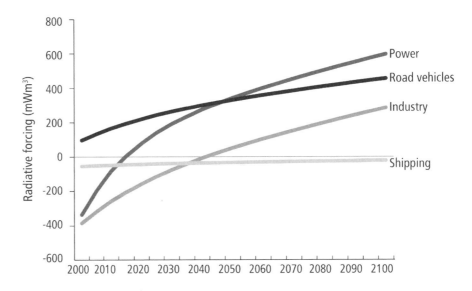

Figure 1.5   A century of total radiative forcing due to constant year 2000 emissions by global sector. (Source: N. Unger, et al., *Proc. Natl. Acad. Sci.*, February 23, 2010, fig. 2.)

The main difference is that on-road transportation is a principal target along the continuum from short to long term, as it is expected to be a top source of climate change throughout this century.

On-road transportation includes cars, light-duty trucks (SUVs, pick-ups, and minivans), medium- and heavy-duty trucks, buses, and motorcycles. The US inventory of on-road transportation direct emissions totaled 1,560 teragrams $CO_{2\,Eq.}$ in 2010.[34] All told, light-duty vehicles—cars, SUVs, pick-ups, and minivans—accounted for about three-quarters of GHG emissions.

It is important to note that ozone is not included because it is not directly emitted; it is formed chemically in the atmosphere by other GHG precursors (principally NMVOC and NO_x) that are emitted directly. On-road transport is responsible for the vast majority (60–95 percent) of total transport GHG emissions, except in the case of SO_2 and N_2O emissions.

The vast majority of these on-road GHG emissions resulted from burning fossil fuels in vehicle engines. On-road vehicles were responsible for about 86 percent of all transportation petroleum use in 2009.[35] Cars and light trucks again dominate oil use, making up more than 75 percent of all on-road vehicles. Motor gasoline is responsible for over one-half of US carbon dioxide emissions from transportation sources.

The differences in ratios of warming effects between transportation sources are approximately the same as the differences in ratios between the carbon dioxide emissions from these sectors. *Thus, it is mainly carbon dioxide that controls the climate response on time scales beyond a decade or two.*[36]

The transportation subsectors—road, air, rail, and shipping—affect the climate on different time scales because of a mix of short-lived and long-lived warming and cooling components. NASA's analysis, as well as research by the Center for International Climate and Environmental Research–Oslo,[37] suggests that motor vehicles (cars, trucks, buses, and motorcycles) are the greatest global contributors to near-term atmospheric warming.[38]

After 100 years, the net temperature change because of global on-road transport is significantly higher than the change stemming from aviation, shipping, and rail.[39] Aviation and shipping have strong but quite uncertain short-lived effects—aviation fuels warming and shipping fuels cooling. Together, these truck and air activities are thought to create net cooling effects during the first decades after the emissions are released.

Dealing with Transportation-Driven Climate Disruption

Climate model outputs have recently been restructured, making it easier to identify emission sources, warming potentials, and temperature increases by economic sector and subsector. This disaggregates the huge IPCC *energy sector*, an amalgamation of disparate activities with massive GHG emissions. These new scientific results make it easier to identify which human activities have the greatest impact on the climate over time. Significant effects of today's GHG emissions will likely be seen over the next century because a carbon dioxide molecule can exist in the atmosphere for 100 years or more. As such, many chemicals involved in climate change will far outlive current US elected officials. Dealing effectively with climate change will require a lasting paradigm shift in public priorities on both a national and an international scale.

Scientists and policy makers must treat air pollution and climate change as related challenges, not two distinct problems. Tightening air-pollution mitigation standards will be a major factor in determining climate-warming trends of the coming decades. Research shows that carbon dioxide and ozone play the most significant roles in near- and mid-term climate forcing.

Reducing climate impacts from on-road transportation will require a shift to new low-carbon fuels. Breakthroughs in the transportation sector at large will come in many forms. In jet aircraft, for example, substitutes for aviation fuels could come from oilseed plants that burn cooler, thus reducing emissions, weight loads, and engine wear. Such innovations could help address mounting concerns in the fastest growing, yet unregulated, subsector—international aviation.[40]

This transition, however, must be accompanied by improved scientific evidence and data keeping on the climate characteristics of each fuel through its entire fuel cycle. Large uncertainties associated with the growing supply of transportation biofuels remain. It has yet to be determined whether new fuels from crops, trees, and other biomass sources actually reduce GHG emissions in practice.

Improvements in battery technology and clean-power generation will lead to higher-confidence estimates of the impact of various technological advances (including a larger plug-in hybrid fleet) on the on-road transportation and power-generation sectors. The simulations conducted by NASA scientists indicate that near-zero-emission electrified transportation may result in a very large reduction in climate forcing, making this technological shift an extremely worthwhile pursuit. The potential to improve the climate and alleviate local pollution justifies increasing investments in electric-vehicle battery research and development as well as clean power.[41]

Despite these promises, trends are headed in a higher-carbon direction. As conventional sources of oil are depleted, unconventional oils with even higher carbon contents will flow at increasing rates through the system to fill the gap. Oil sands from Canada, oil shale from the Rocky Mountain states, and liquefied US coal will only *hyper-carbonize* our transportation system, with far greater amounts of carbon than conventional crude oil. Heavy oils from oils sands and oil shale must be upgraded to a synthetic crude oil to made acceptable at the many refineries that can process only light, conventional crude oils. Upgrading strips out excessive carbon and removes heavy metals and other impurities that are problematic. Partial upgrading renders the heavy oil/bitumen suitable for transportation via pipeline to a refinery for further downstream processing. The contaminants removed from these unconventional oils can pollute water and foul air. US policy makers must carefully investigate and regulate the climate and other environmental impacts of new oil fuels.

Finally, above all, scientists warn that we must avoid mitigation measures that address one problem yet worsen another. This entails identifying unintended consequences of various vehicle and fuel options at the outset. For example, diesel fuel presents a quandary. While diesel engines are more fuel-efficient than gasoline engines, diesel exhaust is also more toxic and contains black carbon that results in short-term climate forcing. Certain biofuels also have potential trade-offs when it comes to their energy and climate effects. Scientists are planning to partner with environmental economists to determine the damage costs (in terms of climate and air quality) from all sectors. These results can be used to develop alternative mitigation scenarios.

Policy-Making Considerations

Technical findings about the relationship between climate and transportation, while extremely valuable, do not consider economics, politics, or social factors. Mitigating on-road transportation climate emissions will not be easy, but this is necessary to reduce risks of climate change.

Moving from scientific knowledge to policy action will require a public-policy-driven paradigm shift in the transportation sector. The market alone cannot accelerate change in this sector, which is dominated by automobiles as well as the oil companies, institutions, land uses, and lifestyles that support them. Americans depend on their cars; in fact, 91 percent of all passenger-miles traveled, excluding air travel, are by car.[42] Petroleum is overwhelmingly the fuel of choice and is used by 94 percent of vehicles; there are few readily available substitutes.

Changing technology and behavior will be most successful if advocates adopt a strategic approach, promoting change over time and embracing paradigm shifts when opportunities arise. This will surely require that a price be placed on transport carbon through a carbon fee, fuel tax, or other incentives and fees. New low-carbon transportation and fuel options can be funded through revenues collected on the most overused, least-efficient, and highest-carbon portions of the transportation system. Once viable travel options have been established, policies can be modified gradually, encouraging wholesale shifts in sector supply and demand. Strategically pricing transportation carbon will be critical to motivating a major paradigm shift for vehicles and oil.

America taxes gasoline less than do most other nations. Only two countries—Kuwait and Saudi Arabia—charge lower gas taxes than the United States and both are net global oil suppliers, not consumers. The federal gas tax has remained unchanged at 18.4¢ for a gallon of gasoline (and 24.4¢ for diesel) for nearly two decades. It is not indexed to the price of crude oil or inflation, so Americans pay a fixed amount whether oil prices are high or low. Taking inflation into account, the gas tax has eroded to only 11¢ today.[43] The gas tax funds a broad range of economy-bolstering transportation projects across the country, and it is already too low to meet current (and future) infrastructure needs.

Transitioning to a lower-carbon transportation system (through plug-in electric vehicles, for example) could result in a substantial benefit for the climate. A technology shift to zero-emission vehicles that results in a 50 percent reduction in on-road transportation emissions is projected to zero-out this source of climate warming. These climate benefits from electrified on-road transportation are expected over both 20- and 100-year time frames. Mitigating climate emissions in on-road transportation, however, will require that electric transportation support both climate and air-pollution goals. As such, electricity will have to be generated by low-carbon renewable and nuclear sources.

Science can provide technical guidance for setting policy priorities. Recent scientific analyses point to on-road transportation as a win-win-win opportunity—good for the climate in the short term and in the long term, and good for our health.[44] Reforming transportation in ways that reduce both emissions and energy consumption is also good for the US economy and can enhance domestic and global security.

US policy makers need to remember four essential factors when developing new strategies. First, on-road transportation is an immediate high-priority target in the short term for reducing greenhouse gas emissions and mitigating climate change. Second, the transportation sector—given its current affinity for oil—is responsible for high levels of long-lived carbon dioxide and ozone precursor emissions that will warm the climate for generations to come. Low-carbon electrification for plug-in vehicles will accelerate the transition to near-zero greenhouse-gas-emission (GHG) cars and trucks. And finally, America's transportation culture must adapt to less reliance on fossil fuels through technological innovation, pricing measures, and sound investments that expand low-carbon mobility choices and fundamentally shift travel behavior.

The climate problems attributed to cars and trucks are not going away. A complementary set of policies will be necessary to simultaneously mitigate global warming, further reduce air pollution, improve vehicle efficiency, and avoid extraction of new, high-carbon oils. Carbon must be priced to realize these potential gains. Rational pricing and wise investments can turn around the US transportation system, bolstering US economic productivity. A clean, efficient, solvent transportation sector will be increasingly vital to the bottom line.

Notes

1. This chapter is largely drawn from: Deborah Gordon, "The Role of Transportation in Driving Climate Disruption" (Carnegie Endowment publication, December 2010).

2. NASA, "Road Transport Emerges as Key Driver of Warming in New Analysis from NASA," February 18, 2010, www.nasa.gov/topics/earth/features/road-transportation.html.

3. The term "climate forcing" is used throughout this report. Climate forcing (also known as "radiative forcing") corresponds to the amount of energy the Earth receives from the sun, and the amount of energy the Earth radiates back into space. A positive forcing (more incoming energy) tends to warm the system, while a negative forcing (more outgoing energy) tends to cool it. Climate-forcing sources include changes in incident solar radiation and changes in concentrations of radiatively active gases (greenhouse gases) and aerosols. Climate forcing is

the initial driver of a climate shift. The thermal inertia of the ocean, land use, and greenhouse-gas and aerosol emissions in the atmosphere determine the speed of climate change.

4. NOAA, "State of the Climate in 2009," *Bulletin of the American Meteorological Society* (*BAMS*) 91 (July 2010), www1.ncdc.noaa.gov/pub/data/cmb/bams-sotc/2009/bams-sotc-2009-chapter2-global-climate-lo-rez.pdf.

5. US Environmental Protection Agency, "Climate Change Science Facts," www.epa.gov/climatechange/downloads/Climate_Change_Science_Facts.pdf.

6. Ibid.

7. World Meteorological Organization, "Current Extreme Weather Events," August 11, 2010, www.wmo.int/pages/mediacentre/news/extremeweathersequence_en.html.

8. National Research Council (NRC), "Climate Stabilization Targets: Emissions, Concentrations, and Impacts Over Decades to Millennia," 2010, books.nap.edu/openbook.php?record_id=12877&page=1.

9. Ibid.

10. NRC, ibid. Note: 1,000 GtC is equivalent to 3,666 gigatonnes carbon dioxide. In order to convert mass units, 1 gigatonne (Gt) = 1 billion metric tons = 1,000 teragrams (Tg). Note that in 2011, 31.6 Gt CO_2 were emitted, which is fast approaching the 2017 32.6 Gt level the IEA has determined is consistent with a 50 percent chance of limiting an average global temperature increase to 2°C. For more information see: *Clean Technica* (*http://s.tt/1cJk2*).

11. These researchers include: Center for Climate Systems Research at Columbia University; University of Illinois, Urbana-Champaign; Lawrence Berkeley National Laboratory; and Environmental Defense Fund.

12. Land-use emissions include land use, land-use change, and forestry activities. Land use serves as a carbon dioxide source and sink, resulting in a flux that includes both emissions and sequestration. In the United States most land-use emissions are carbon sinks.

13. The following data source was used extensively in this section: EPA, "Inventory of US Greenhouse Gas Emissions and Sinks: 1990–2010," EPA 430-R-12-001, April 15, 2012, www.epa.gov/climatechange/emissions/downloads12/US-GHG-Inventory–2012-ES.pdf. Note that the total mass of GHG emissions is an *indicator* but not a *predictor* of climate impacts. Climate models take into account complex chemical reactions of direct emissions and air pollutant precursors in order to predict radiative forcing from emission inputs.

14. Ibid., table ES–3.

15. Precursor emissions are compounds that react to form yet other air pollutants, such as smog. These also serve as the building blocks of climate gases and other toxic air contaminants.

16. UNFCCC reporting requirements on indirect greenhouse gases to date include: CO, NO_x, NMVOCs, and SO_2, and are reviewed in unfccc.int/resource/docs/cop8/08.pdf. PM, BC, and aerosols are not yet reported per UNFCCC guidelines.

17. SO_2 is an indirect GHG in its own right and also forms aerosols (fine particles) along with particulate matter. The transportation sector is not a significant producer of aerosols,

compared to industry and electric utilities burning coal. However, heavy-duty trucks (burning diesel), ships, and off-road transport equipment are the sector's main aerosol (SO_2 and PM) contributors.

18. Scientific studies link breathing PM to significant health problems, including aggravated asthma, difficult breathing, chronic bronchitis, myocardial infarction (heart attacks), and premature death. Diesel exhaust is likely a human carcinogen by inhalation and poses a non-cancerous respiratory hazard. PM is also the major source of haze that reduces visibility, and can cause the erosion of structures such as monuments and statues. Particulate matter generated by fuel combustion (such as diesel engines) tends to be smaller on average than particulate matter caused by sources such as windblown dust.

19. In developing nations, the burning of organic fuels (animal waste, trees, grasses) contributes significant BC emissions.

20. See: EPA, "Linkages Between Climate and Air Quality," August 30, 2010, www.epa.gov/AMD/Climate.

21. For more energy and transportation information, see: Oak Ridge National Laboratory, "Transportation Energy Databook, Edition: 30," June 2011, cta.ornl.gov/data/download30.shtml.

22. Ibid. Note: the alcohol fuels (corn ethanol) blended into gasoline to make gasohol (10 percent ethanol or less) are counted under "renewables" and are not included in petroleum share.

23. There is a small portion of the fuel that is not oxidized into carbon dioxide when the fuel is burned. EPA has published information on carbon dioxide emissions from gasoline and diesel, taking the oxidation factor into account based on the carbon content used in EPA's fuel economy analysis.

24. US EPA, "Emission Facts: Average Carbon Dioxide Emissions Resulting from Gasoline and Diesel Fuel," February 2009; additional resources available at www.epa.gov/OMS.

25. Based on calculations from oil sands compositions cited in: O. P. Strausz, "The Chemistry of the Alberta Oil Sand Bitumen" (Hydrocarbon Research Center, Department of Chemistry, University of Alberta, Edmonton, Alberta, Canada), www.anl.gov/PCS/acsfuel/preprint%20archive/Files/22_3_MONTREAL_06–77_0171.pdf.

26. S. Unnasch et al. (Life Cycle Associates), "CRC Report No. E–88: Review of Transportation Fuel Life Cycle Analysis" (report prepared for Coordinating Research Council Project E–88, 2011); and S. Unnasch et al., "Assessment of Life Cycle GHG Emissions Associated with Petroleum Fuels," Life Cycle Associates Report LCA–6004–3P (report prepared for New Fuels Alliance, 2009), www.newfuelsalliance.org/NFA_PImpacts_v35.pdf.

27. See 2007 Tiax report for the California Air Resources Board and California Energy Commission on the GREET model, www.arb.ca.gov/fuels/lcfs/greet.pdf.

28. Adam R. Brandt, "Variability and Uncertainty in Life-Cycle Assessment Models for Greenhouse-Gas Emissions from Canadian Oil Sands Production," *Environmental Science*

& Technology (2012) doi: 10.1021/es202312p, summary available at www.greencarcongress .com/2012/01/brandt–20120106.html.

29. For example, see: Worldwatch Institute, 2009, www.worldwatch.org/node/6082.

30. This section largely drawn from publications about Nadine Unger's findings: www .pnas.org/content/early/2010/02/02/0906548107.full.pdf+html and www.energybulletin.net /node/51744.

31. The indirect effect of sulfur-derived aerosols on radiative forcing can be considered in two parts: (1) the aerosols' tendency to decrease water droplet size and increase water droplet concentration in the atmosphere; and (2) the tendency of the reduction in cloud droplet size to affect precipitation by increasing cloud lifetime and thickness. Although still highly uncertain, the radiative forcing estimates from both effects are believed to be negative. Since SO_2 is short-lived and unevenly distributed in the atmosphere, its radiative forcing impacts are highly uncertain. Unger notes that NASA's Glory mission may help reduce the uncertainties associated with aerosols.

In addition to acting as an indirect GHG, SO_2 is also a major contributor to the formation of regional haze, which can cause significant increases in acute and chronic respiratory disease. For this reason, SO_2 is a criteria pollutant under the Clean Air Act. Once SO_2 is emitted, it is chemically transformed in the atmosphere and returns to the Earth as the primary source of acid rain. Electricity generation (using coal) is the largest anthropogenic source of SO_2 emissions in the United States, accounting for 80 percent. SO_2 emissions have decreased in recent years, primarily due to power generators switching from high-sulfur to low-sulfur coal and installing flue-gas desulfurization equipment. While beneficial for air quality and acid rain mitigation, the removal of SO_2 aerosols has inadvertently also meant removing cooling agents from the climate, thus accelerating warming trends.

32. Nadine Unger, NASA blog, December 4, 2009, climate.nasa.gov/blogs/index.cfm?Fuse Action=ShowBlug&NewsID=139.

33. NASA, *Earth Science News*, February 18, 2010, www.nasa.gov/topics/earth/features /unger-qa.html.

34. EPA, *Trends in GHG Emissions*; see table 2–15.

35. ORNL, Transportation Energy Databook; see table 1.15.

36. Jens Borken-Kleefeld, T. Berntsen, and J. Fuglestvedt, "Specific Climate Impacts of Passenger and Freight Transport," *Environmental Science & Technology* 44, no. 15 (July 12, 2010), pubs.acs.org/doi/pdfplus/10.1021/es9039693.

37. CICERO is the Center for International Climate and Environmental Research–Oslo (Norway).

38. Aviation has a very high climate-forcing contribution from short-lived contrails and cirrus clouds that quickly decrease with time. Still, aggregated on-road transport emissions are greater than those of air travel in the mid- to long term. See: Borken-Kleefeld, Berntsen, and Fuglestvedt, "Specific Climate Impacts."

39. Jan Fuglestvedt, Terje Berntsen, Gunnar Myhre, Kristin Rypdal, and Ragnhild Bieltvedt Skeie, "Climate Forcing from the Transport Sectors," *PNAS* 105, no. 2 (January 15, 2008): 454–58, fig. 2A, www.pnas.org/content/105/2/454.full.

40. The European Union (EU) has decided to cap such emissions and make them part of the EU Emissions Trading System by 2012. The first international flights using advanced biofuels (e.g., Camelina and Jathropha-curcas) occurred in July 2011. There are trade-offs, however. Rapid conversion to advanced aviation biofuels faces a lack of sufficient feedstock and refinery capacity, resulting in increased cost at the margin. In August 2011 the US Departments of Energy and Agriculture, and the US Navy, announced a $10 million, three-year partnership to produce drop-in aviation and marine biofuels in collaboration with private industry. Separately, the US Air Force has set a goal of sourcing 50 percent of its aviation fuel from renewable resources by 2050.

41. Nadine Unger, "Transportation Pollution and Global Warming," June 2009, www.giss.nasa.gov/research/briefs/unger_02/; for full article see: N. Unger, D. T. Shindell, and J. S. Wang, "Climate Forcing by the On-Road Transportation and Power Generation Sectors," *Atmospheric Environment* 43 (2009): 3077–85. Note that the additional electricity needed to power EVs, at least in the short term, would likely be generated by coal, thus further reducing global warming through increased emissions of cooling sulfate particles. This is not a desirable long-term solution, of course, as the power sector emits huge quantities of CO_2 along with smog, particles, and other pollutants that are detrimental to public health. However, in the short term, an electrified on-road transportation system would be a boon to climate mitigation.

42. When air PMT is included, autos still account for a whopping 81 percent of total US PMT. Total 2008 travel equaled 5.5 billion PMT. See: Bureau of Transportation Statistics, www.tbs.gov/publications/national_transportation_statistics/html/table_01_37.html, table 1–37.

43. Shin-Pei Tsay and Deborah Gordon, "Five Myths About Your Gasoline Tax," CNN.com, November 19, 2011, www.cnn.com/2011/11/18/opinion/tsay-gordon-gas-tax-myths/index.html.

44. For hundreds of thousands of people, smog-polluted (and soot-filled) air means more breathing problems and aggravated asthma, trips to the emergency room, and hospital admissions, particularly to intensive-care units. Moreover, studies have shown that ozone pollution at current US levels contributes to early death. See: M. L. Bell, F. Dominici and J. M. Samet, "A Meta-Analysis of Time-Series Studies of Ozone and Mortality with Comparison to the National Morbidity, Mortality, and Air Pollution Study," *Epidemiology* 16 (2005): 436–45; J. I. Levy, S. M. Chermerynski, and J. A. Sarnat, "Ozone Exposure and Mortality: An Empiric Bayes Metaregression Analysis," *Epidemiology* 16 (2005): 458–68; K. Ito, S. F. De Leon, and M. Lippmann, "Associations Between Ozone and Daily Mortality: Analysis and Meta-Analysis," *Epidemiology* 16 (2005): 446–29; D. V. Bates, "Ambient Ozone and Mortality," *Epidemiology* 16 (2005): 427–29.

2 Oil Vulnerability in the American City

NEIL SIPE AND JAGO DODSON

This chapter investigates the varying intersection of volatile petroleum markets and housing finance pressures with household socioeconomic status and urban structure, using six American cities (Atlanta, Boston, Chicago, Las Vegas, Phoenix, and Portland) as case studies. The chapter responds to two important economic phenomena seen over the past several years. The first is the sharp growth and volatility in global petroleum prices between 2004 and 2012, which mark a dramatic departure from predominantly stable and low prices seen since World War II. The result of the post–2004 oil price gains has been marked gains in the cost of fuel in most nations, which in turn have raised concerns about the effect of higher fuel prices on the household sector. Oil prices remain extremely volatile—rising from around $40 per barrel in 2004 to over $145 in mid–2008 before dropping to $30 per barrel in 2009 and then rising again to over $100 in 2011 and finally falling to $85 per barrel, where it sits today. The result has been higher gasoline prices in US cities of around $4 per gallon.

The second phenomenon is the growth in house prices in many urban housing markets, high levels of household housing debt, and the subsequent declines in these markets generated by the global financial crisis. Since late 2007, global credit markets have experienced a set of failures that are, in part, tied to home ownership lending, especially in the sub-prime sector. This problem has impacted urban housing markets in many nations. While governments have rushed to reduce official interest rates to avert wider financial problems, the household sector is still facing considerable stress from high housing debt and ongoing doubts about the magnitude of the global financial crisis.

The recent problems posed by volatile fuel prices in combination with high levels of housing debt raise important questions for urban systems. Transport systems in the majority of US cities are highly car-dependent and are, in fact, the most car-dependent national group in the world.[1] Higher and volatile urban fuel prices raise the financial risk and cost of car-dependent transport and add to household financial stress. In many cities, owner-occupied housing tenure is intimately tied to car-based suburbanization, although this tenure-transport link has been underexplored.[2] We argue that housing systems, which have been spatially structured on the basis of relatively cheap transport fuel, face a process of difficult adjustment in a volatile and likely higher fuel-cost environment that will intersect with global economic weakness, especially in credit markets. In this changing energy and financial context, the question of whether uncertainty over the cost of transport fuel will undermine the historic links between car-based suburban transport and tenure patterns deserves serious attention. In turn, we suggest there is a need for urban social science to better understand these transport and housing patterns, impacts, and effects.

The volatility in global petroleum prices raises new questions about contemporary urban transport systems and in turn about the form and tenure of the housing arrangements they support. Likewise, the partial basis of the current financial crisis in housing markets poses further questions about socio-tenure structures and spatial patterns within cities. Such questions will have to be answered in an increasingly multifaceted environment for city-regions that are being transformed by ongoing economic and social reconfigurations as well as flows of capital and labor.[3] These questions are likely to be particularly acute in global city regions—including "post-suburban" regions[4] that are experiencing increasingly complex governance,[5] spatial,[6] economic,[7] social,[8] demographic, and technological[9] changes.

In this fluid context, the combination of household oil vulnerability and household mortgage stress could pose a considerable social and economic challenge for governments. The economic and social effects on cities of the newly volatile global petroleum context and recent financial dysfunction deserve serious attention from scholars, as these factors appear set to play a new and forceful role in the development of advanced city-regions. There is a need to examine the effects of household energy and financial stress on major global metropolitan systems as well as the processes of change that they portend.

Urban literature that considers the relationship between housing tenure and transport systems from a spatial perspective is sparse. Few papers have specifically addressed the relationship between transport and housing tenure, and it is rarely discussed in transport or housing debates. Krizek's[10] and the Center for Neighborhood

Technology's[11] work on "location efficient" mortgages are among the few analyses in this area, beyond classic work on urban structure, bid-rents and transport,[12] analyses of "spatial mismatch,"[13] or Kemeny's[14] theoretical work on the links between political economy, owner occupation, and urban form. There has been little recent research interaction between housing sociologists and transport geographers. This relative gap in the research record is surprising, given that the history of modern urbanization shows that transportation has been a crucial means for solving housing problems.[15] During the early twentieth century, transport networks were used to address the housing problems in many crowded industrial cities by providing access to cheap, less-developed exurban land. While suburbanization processes took markedly different forms across different jurisdictions and tenure systems, the use of transport to solve housing problems was a hallmark of twentieth-century urban modernization.

Because there are few studies that have considered these issues, we have only a modest understanding of how these processes will intersect in cities, especially in the large spatially complex metropolitan areas that have emerged in recent decades.[16] Dodson and Sipe began charting the potential spatial distributional consequences of these issues in Australia but did not offer a prospective diagnosis.[17] It is critical that scholars begin to explore and investigate the contours of these patterns in order to understand the dynamics of energy and financial stress on urban socioeconomic structures and patterns. This task will inevitably need to draw on the considerable body of work on social polarization within cities that has been reported over the past two decades. This effort will also require new methods for evaluation that combine information about different factors that are involved in these new processes.

The remainder of this chapter investigates the social consequences of declining petroleum security and increased financial instability on six American cities as case studies. The chapter has three primary objectives. First, we use the case studies to identify how declining petroleum security and weakening household financial capacities intersect at the urban scale. Second, we examine the intersection of household socio-tenure stress patterns with other urban variables. A final objective is to ask what will be the broader urban challenges faced by global city-regions in dealing with volatile fuel prices and the household effects of the global financial crisis in light of an emerging literature on urban energy security. The chapter's conclusions reflect upon the role of energy and finance in shaping contemporary urban patterns and offer some policy suggestions for reducing the spread of socio-spatial energy and financial risk in cities. The chapter ends with a call for renewed effort to improve social scientific understanding of the links between energy disadvantage, fuel poverty, transport, and housing systems.

Petroleum, Transport, and Urban Structure

The global price of oil was barely more than $30 per barrel at the start of 2004 but grew rapidly to reach $140 per barrel by mid–2008. At the time of writing, global oil prices had subsided to approximately $85 per barrel but remained volatile as a result of ongoing uncertainty in global financial markets. There are multiple reasons behind the growth and volatility in the price of oil, which we touch on only briefly here. They include: rapid global economic growth, production constraints and disruptions, geopolitical tensions, and longer-term fears about resource sustainability. Such problems and constraints have been canvassed widely elsewhere[18] as well as in this book (see chap. 1). The longer-term prospects for low petroleum prices appear to be deteriorating. The International Energy Agency's 2008 *World Energy Outlook* estimates that oil prices will return to the high levels seen in 2008 within a decade, and the agency's chief economic analyst, Fatih Birol, has admitted that "we are expecting that in three, four years' time the production of conventional oil will come to a plateau, and start to decline."[19]

The shifts in global petroleum conditions during 2004–2008 translated into much higher gasoline prices in American cities, at times in excess of $4 per gallon.

In American cities, the urban socioeconomic impacts of higher gasoline prices are differentiated spatially as a result of urban structure. In general, the inner metropolitan parts of older American large cities developed around historical public transport routes and exhibit a higher-density urban form. By contrast, suburban areas beyond the central business district, especially those developed after World War II, tend to have a low-density urban form and are dependent on private motor vehicles for transport. The result of this post-WWII shift is that collectively, America's cities, including the six examined in this chapter, represent the most car-dependent urban regions in the world.[20]

The structure of transport systems, housing markets, housing tenure, and socioeconomic status at the level of the city-region are likely to be powerful influences on the way that households experience volatile transport-fuel costs and housing-finance difficulties. Many urban regions are already marked by profound social differences mediated by infrastructure networks, housing systems, and the distribution of economic and other opportunities.[21]

Investigating the Structure of Transport, Tenure, and Socioeconomic Status in Cities

This chapter adapts a method we initially developed to investigate cross-sectional patterns of petroleum and mortgage vulnerability in Australian cities.[22] The method has been adapted slightly to account for shifts in the way the US Census Bureau reports their data. This chapter uses that approach to understand patterns of petroleum and mortgage vulnerability in selected American cities—Boston, Chicago, Las Vegas, Phoenix, Atlanta, and Portland. The remainder of this chapter reports the results of analysis of the distribution of household petroleum and mortgage vulnerability in these cities at the household level in 2000 and then some of the factors influencing these patterns.

These six cities or Metropolitan Statistical Areas (MSAs) were chosen to illustrate the index and were not intended to be representative of all US cities, although we did ground our selection on some general criteria. The MSAs were chosen on the basis of such factors as age, population growth, availability of public transport, and broader urban structure. Boston and Chicago were chosen because they are older cities that developed prior to the motorcar and had initially developed alongside heavy-rail public transit. Atlanta, Las Vegas, and Phoenix were chosen because of their rapid growth and relatively younger age, which has resulted in development along roads rather than alongside fixed-rail transit. Finally, Portland was chosen because it has attained a high profile for taking steps to reduce auto dependence by building a light-rail system and implementing an urban growth boundary. Thus, these cities provide sufficient variety to test the Oil and Mortgage Vulnerability Index.

The method developed in this chapter is comparable to an index developed separately by the Center for Neighbourhood Technology (CNT) to assess housing and transport costs at the individual household scale. This index grew, in part, out of the location-efficient mortgages concept that was promoted by the CNT.[23] The CNT index uses nine variables—six neighborhood variables (residential density, gross density, average block size, transit connectivity index, job density, and average journey-to-work time) and three household variables (household income, household size, and commuters per household). With these nine factors they estimate total transportation costs resulting from car ownership/use and public transit usage.

The CNT index is the closest measure to our Oil and Mortgage Vulnerability Index but is considerably more complicated in terms of data requirements, assumptions, and preparation time. One reason for the increased complexity of the CNT index is that it looks at actual transport, housing, and energy costs, while our index measures levels of vulnerability, not direct economic impacts. The CNT index measures affordability, while our index measures vulnerability. Nevertheless, the two methods examine the same issue—the distribution of the combined impact of housing and transportation costs relative to income across an urban area. While we have not done a side-by-side comparison of the results from the two indices, a cursory examination suggests the conclusions are the same: large areas of US cities are vulnerable due to increased car dependence. The goal of our effort is to highlight the vulnerability problem and elevate its importance to policy makers. We believe that a methodology that is simple to understand and quick to prepare, and that can be generalized between localities and cities, is an important part of achieving policy purchase—a key concern for planners with an interest in sustainable urban outcomes.

Assessing Spatial Petroleum and Mortgage Vulnerability

We originally developed a GIS-based method for investigating the extent of household exposure to higher petroleum and mortgage debts using Australian census.[24] The method has been adapted for US cities by constructing the vulnerability methodology using four variables from the 2000 US Census. These four variables are combined to provide a composite vulnerability index that can be mapped at the census block-group level. The block group was used because it is the smallest geographical unit for which the US Census Bureau publishes sample data. This Oil and Mortgage Vulnerability Index assesses the average vulnerability of households within the block group rather than the specific vulnerability of individual households. The variables used are

> *Car dependence:*
> • Proportion of working individuals who took a journey to work by car (either as a driver or passenger)
> • Proportion of households with more than two cars
> *Mortgage tenure:*
> • Proportion of dwelling units that are being purchased
> *Income level:*
> • Median household income

The journey-to-work variable indicates the basic level of demand for automobile travel, while the household rate of vehicle ownership indicates the extent of household investment in motor-vehicle travel. Together these two variables provide an indicator of car dependence and of relative household-demand exposure to volatile and rising costs of motor-vehicle travel.

The mortgage variable represents the prevalence of mortgage tenure and, accordingly, household exposure to interest rate rises. It is important to note that it does not incorporate housing costs directly. The income variable is used to register the financial capacity of households within the locality to adjust to fuel and general price volatility. Together these four variables provide a basic, but comprehensive, spatial representation of household mortgage and oil vulnerability.

The Oil and Mortgage Vulnerability Index was constructed by combining the four variables (as shown in table 2.1). Higher levels of car ownership, journey to work by car, and mortgage tenure received higher index values, while low levels of household incomes received lower scores. Thus a block group with high levels of car ownership, journeys to work by car, mortgages, and low incomes would receive a score of 20 (5+5+5+5). However, the four variables we have selected are not equal in their contribution to the Oil and Mortgage Vulnerability Index and have been given weightings as shown in table 2.2. Thus, of a total possible index score of 30, five points are provided by each of the car ownership and journey-to-work variables, while ten points each are provided by the income and mortgage scores. Thus car dependence comprises one-third of the index, the proportion of homes with mortgages another one-third of the index, and median household income making up the remaining one-third. We have reported these weightings to be robust values.[25]

Table 2.1 Value assignment relative to Census Block Group percentile for the Oil and Mortgage Vulnerability Index

	Value assigned			
	> 2 Cars	JTW by car	Income	Mortgage
100	5	5	0	5
90	4	4	1	4
75	3	3	2	3
50	2	2	3	2
25	1	1	4	1
10	0	0	5	0

Table 2.2 Variable weighting for the Oil and Mortgage Vulnerability Index

Indicator	Proportion of households with >two cars	Proportion of work trips by car	Median household income	Proportion of households with a mortgage
Maximum potential points	5	5	10	10
Weighting	33.3%		33.3%	33.3%

Results

The results of the analysis (see table 2.3 and fig. 2.1) show that Boston and Chicago had the lowest percentage of block groups in the high and very high categories, with 19.9 percent and 21.7 percent, respectively. Portland was slightly higher with 24.8 percent in the high categories. The remaining three cities examined had more than double the percentage of block groups in the high categories, with Atlanta at 51.7 percent, Las Vegas at 50.6 percent, and Phoenix 49.7 at percent.

Maps for each of the six MSAs showing the distribution of oil and mortgage vulnerability are provided in figures 2.2–2.7. The goal of these maps is to show general patterns of vulnerability at the metropolitan level, not to examine oil vulnerability in specific block groups.

To maintain consistency between the metropolitan areas, they are all shown at the same scale (1" = 5 miles) and use the same categories for vulnerability levels. The division between categories on each map is based on percentiles and are the same as those used for each variable in the index (0–10, 10–25, 25–50, 50–75, 75–90, and 90–100). To keep the maps from becoming cluttered, only freeways are shown as points of reference. We acknowledge that it would have been useful to include major transit corridors as well, but that was beyond the scope of this analysis. Finally, on some of the

Table 2.3 Oil and mortgage vulnerability values by block group

		Oil and Mortgage Vulnerability Index					
		Low		Moderate		High	
MSA		0–6	6–10	10–16	16–22	22–26	26–29
Atlanta	#	0	9	43	837	936	12
	%	0.0	0.5	2.3	45.6	51.0	0.7
Boston	#	1	825	302	2,775	770	14
	%	0.0	2.1	7.7	70.4	19.5	0.4
Chicago	#	6	118	368	4,244	1,295	18
	%	0.1	2.0	6.1	70.2	21.4	0.3
Las Vegas	#	1	16	45	360	425	8
	%	0.1	1.9	5.3	42.1	49.7	0.9
Phoenix	#	0	35	131	957	1,063	44
	%	0.0	1.6	5.9	42.9	47.7	2.0
Portland	#	0	5	28	1,072	361	3
	%	0.0	0.3	1.9	73.0	24.6	0.2

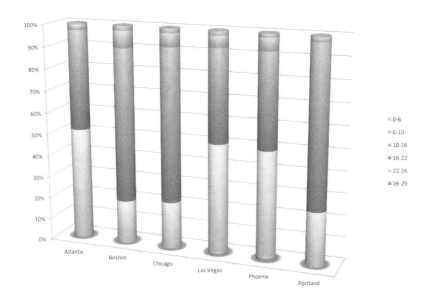

Figure 2.1   Oil and mortgage vulnerability comparison across MSAs.

maps there are large (low-vulnerability) block groups that may have either small or institutional populations.

These results are not unexpected, given that Chicago and Boston are older MSAs that grew and developed around extensive public transit networks. Comparatively, Portland has a much newer public transit network, but their efforts (including the urban growth boundary and other planning initiatives) over the past several decades appear to be having an impact. Portland's level of vulnerability is much closer to that of the older transit-based cities than to Atlanta, Phoenix, or Las Vegas. The results suggest a link between vulnerability and public transit, which deserves more attention than what is provided here, using spatial statistics to investigate the relationship between vulnerability and access to public transit.

Conclusions: Petroleum and Housing-Tenure Vulnerability in Cities

The research we have undertaken demonstrates that there is a broad spatial distribution of socioeconomic exposure to oil and mortgage costs within and across these six MSAs. In general, higher and lower vulnerability are concentrated in different subregions within each MSA. Despite some local variation, higher vulnerability tends to be found in the outer suburban areas (see figs. 2.2–2.7), where cheaper housing attracts modest-income home purchasers and where the main mode of travel is by

Figure 2.2   Atlanta MSA Oil and Mortgage Vulnerability Index.

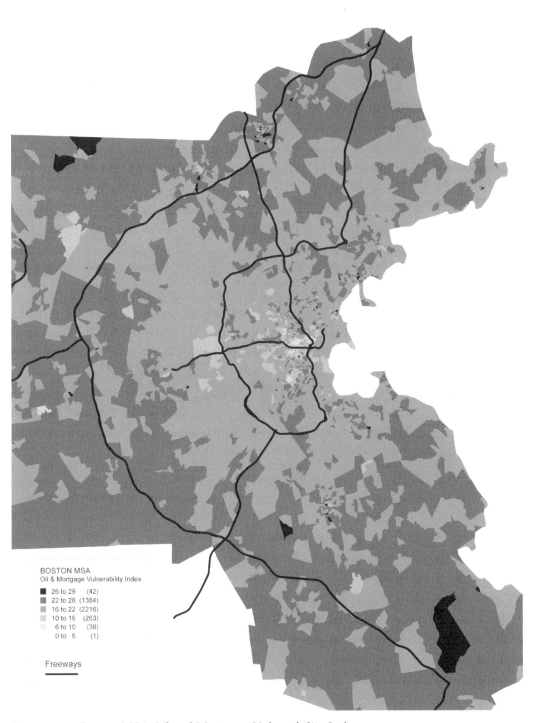

Figure 2.3   Boston MSA Oil and Mortgage Vulnerability Index.

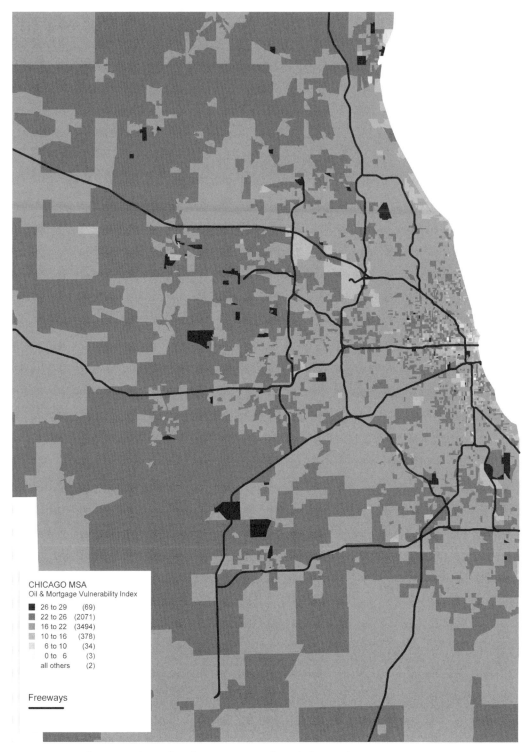

CHICAGO MSA
Oil & Mortgage Vulnerability Index
■ 26 to 29   (69)
■ 22 to 26   (2071)
■ 16 to 22   (3494)
■ 10 to 16   (378)
□ 6 to 10   (34)
  0 to 6   (3)
  all others (2)

Freeways
——————

Figure 2.4   Chicago MSA Oil and Mortgage Vulnerability Index.

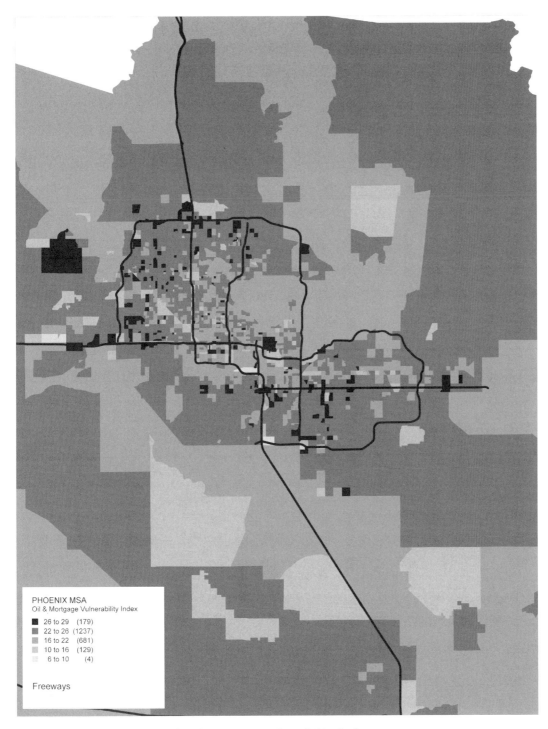

Figure 2.5   Phoenix MSA Oil and Mortgage Vulnerability Index.

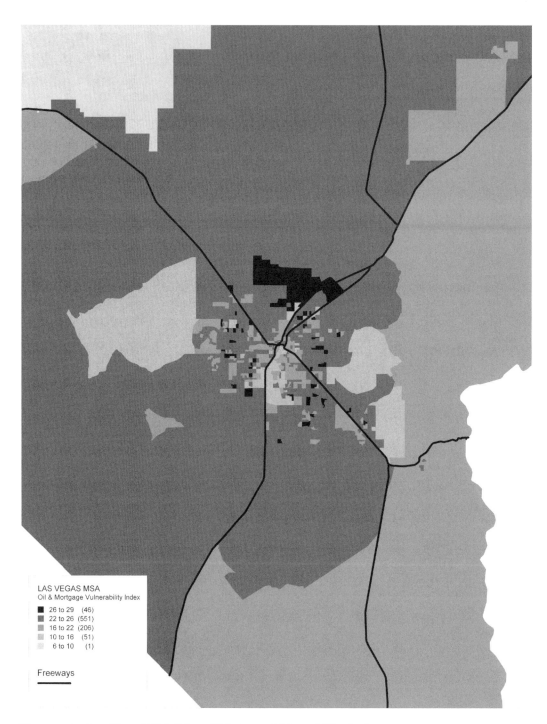

Figure 2.6   Las Vegas MSA Oil and Mortgage Vulnerability Index.

Figure 2.7   Portland MSA Oil and Mortgage Vulnerability Index.

automobile. By comparison, households in inner suburban locations (particularly in Boston, Chicago, and Portland) typically experience the advantages, from a petroleum and mortgage vulnerability perspective, of higher incomes and lower reliance on automobiles for transport than those in outer suburban zones. Urban structure and the local conditions of resilience and adaptability that urban structure engenders will be a critical factor that shapes household socioeconomic circumstances under conditions of volatile petroleum conditions and housing finance uncertainty.

Because this methodology measures spatial *vulnerability* to these pressures, rather than direct impacts, the relative distributional patterns we have described remain relevant whether oil is at $40 per barrel or $160. The recent formal prediction by the International Energy Agency,[26] that petroleum prices will approach or surpass their 2008 highs within a decade, implies the continuing significance of such issues for many American cities as well as comparable cities in Canada, the United Kingdom, Australia, and New Zealand, and, from a broader comparative perspective, in other jurisdictions such as Europe and Asia.

The Regressive City Region

Because the impacts of volatile petroleum prices and household housing debts are borne unevenly across cities, petroleum and mortgage vulnerability is a considerable economic and social challenge, as this example has shown. Within the typical American metropolitan area, exemplified by these six MSAs, the households that will face the greatest task in coping with higher transport and housing costs are those with the most limited financial resources, the greatest automobile and mortgage-finance dependence, and the weakest access to alternative infrastructure such as public transport. Low-income suburban households are also less likely than the wealthier households in middle and inner zones to be able to afford high-priced alternative vehicle types or fuels. The Chevrolet Volt electric car, for example, is expected to retail at approximately $47,000—well above the level that is affordable to low- and moderate-income households. Under current conditions where fuel and housing markets intersect with differentiated tenures and local infrastructure deficits, uncertainty in fuel prices and mortgage-interest rates have resulted in highly regressive conditions. This is likely to be exacerbated by recent processes of "reurbanization" reported in some US cities, whereby higher-income households seek inner-urban locations with numerous amenities, including walkable streetscapes and good access to transit. Such trends may thus concentrate advantaged households in higher-value, less oil-vulnerable locations while displacing less-affluent households to higher-vulnerability sites where housing values are depressed.

Active individual household choice, such as the ambition of owner occupation, plays only a partial role in shaping household locational decisions in American cities. Additional factors, such as the structure of housing and labor markets, strongly limit where households are able to afford to purchase housing. Another factor is the US mortgage subsidy, which encourages overconsumption of housing, particularly among moderate- and high-income households. These factors, including reurbanization, presently allocate less-wealthy households to parts of the city where they face grossly constrained travel choices, resulting in higher car-dependence and thus higher exposure to the socioeconomic risks and uncertainty from volatile fuel prices. Fuel price volatility and housing financial structures are therefore likely to act as deeply regressive features of American urban processes. These patterns are superimposed upon an already inequitable social landscape.[27]

Policy Issues

The patterns described in this chapter raise considerable questions for urban policy, urban governance, and infrastructure investment and provision. Because the impacts of volatile fuel prices and the household effects of mortgage stress are inequitably distributed within cities, any policies to address the socioeconomic stress of changing petroleum conditions and to manage financial instability must account for these structural factors. Governments have a responsibility to redress spatial failures of provision and barriers of social access within the urban and suburban infrastructure landscape—a phenomenon Graham and Marvin describe as "splintering urbanism"[28]—over which they exert the greatest control, such as the distribution of high-quality public transport infrastructure and services.

If American metropolitan areas are to remain socioeconomically resilient in the coming decades, urban planning will need to give much greater emphasis to less car-dependent modes of travel, such as public transit, walking, and cycling. Recent planning in many American cities has made little substantive change in response to these issues though. However, there are a few examples of cities that have begun to address such concerns—Portland, one of the cities examined in this chapter, has initiated policy development on petroleum volatility. It remains to be seen whether higher oil prices will spur new investment in highly oil-dependent regions. Much of the current US strategic policy appears set on achieving greater national energy independence at potentially high environmental cost via access to unconventional petroleum sources, rather than addressing the basic demand factors underpinning oil dependence. However, it is important to acknowledge the billions of dollars of government investment in new light rail, commuter rail, and streetcar systems throughout the United States.

This investment is having a major impact on the reurbanization of cities, which should also help to reduce oil and mortgage vulnerability.

New Landscapes for Urban Research

While the trajectory of global petroleum markets remains uncertain, the mounting expectation of authoritative formal institutions such as the International Energy Agency is for encroaching constraints on petroleum production over the longer term. Such constraints will inevitably intersect with the socio-spatial patterns described in this chapter. Furthermore, the global financial crisis, and its roots in particular housing sub-markets, implies considerable uncertainty both in housing markets and associated financial systems. New questions of socioeconomic polarization and urban spatial inequality posed by these combined factors may therefore be expected to grow in significance in coming years. There is a need for urban scholarship to give greater attention to understanding petroleum, transport, tenure, and debt issues and the intersection of effects on major city-regions. This chapter has contributed some preliminary insights to the urgent and continuing task of responding to these imperatives.

Notes

1. P. Newman and J. Kenworthy, *Sustainability and Cities: Overcoming Automobile Dependence* (Washington, DC: Island Press, 1999).

2. See: Center for Neighborhood Technology's Housing + Transportation Index: htaindex .cnt.org/.

3. A. Scott, "Resurgent Metropolis: Economy, Society, and Urbanization in an Interconnected World," *International Journal of Urban and Regional Research* 32, no. 3 (2008): 548–64.

4. N. A. Phelps and A. M. Wood, "The New Post-Suburban Politics?" *Urban Studies* 48, no. 12 (2011): 2591–610.

5. See: A. Kearns and R. Paddison, "New Challenges for Urban Governance," *Urban Studies* 37, nos. 5, 6 (2000): 845–50; and C. Mitchell-Weaver, D. Miller, et al., "Multilevel Governance and Metropolitan Regionalism in the USA," *Urban Studies* 37, nos. 5, 6 (2000): 851–76.

6. P. Healey, *Urban Complexity and Spatial Strategies: A Relational Planning for Our Times* (London: Routledge, 2006).

7. Scott, "Resurgent Metropolis," 548–64.

8. R. A. Walks, "The Social Ecology of the Post-Fordist/Global City? Economic Restructuring and Socio-Spatial Polarisation in the Toronto Urban Region," *Urban Studies* 38, no. 3 (2001): 407–47.

9. S. Graham and S. Marvin, *Splintering Urbanism: Networked Infrastructures, Technological Mobilities, and the Urban Condition* (London and New York: Routledge, 2001).

10. K. Krizek, "Transit Supportive Home Loans: Theory, Application, and Prospects for Smart Growth," *Housing Policy Debate* 14, no. 4 (2003): 657–77.

11. See: Center for Neighborhood Technology.

12. See: R. F. Muth, "The Spatial Structure of the Urban Housing Market," *Papers and Proceedings of the Regional Science Association* 7 (1961): 207–20; L. Wingo, *Transportation and Urban Land* (Washington, DC: Resources for the Future, 1961); W. Alonso, *Location and Land Use: Toward a General Theory of Land Rent* (Cambridge, MA: Harvard University Press, 1964); and E. S. Mills, "An Aggregative Model of Resource Allocation in a Metropolitan Area," *American Economic Review* 57, no. 2 (1969): 197–210.

13. See: J. Kain, "Housing Segregation, Negro Employment, and Metropolitan Decentralization," *Quarterly Journal of Economics* 82, no. 2 (1968): 175–92; and K. R. Ihlanfeldt and D. L. Sjoquist, "The Spatial Mismatch Hypothesis: A Review of Recent Studies and Their Implications for Welfare Reform," *Housing Policy Debate* 9, no. 4 (1998): 849–92.

14. J. Kemeny, "Home Ownership and Privatisation," *International Journal of Urban and Regional Research* 4, no. 3 (1980): 372–88; and J. Kemeny, "'The Really Big Trade-off' between Home Ownership and Welfare: Castles' Evaluation of the 1980 Thesis, and a Reformulation 25 Years On," *Housing Theory and Society* 22, no. 2 (2005): 59–75.

15. K. Jackson, *Crabgrass Frontier: The Suburbanization of the United States* (Boston, MA: Oxford University Press, 1987); R. Fishman, *Bourgeois Utopias: The Rise and Fall of Suburbia* (New York: Basic Books, 1989); R. Cervero, *The Transit Metropolis: A Global Inquiry* (Washington, DC: Island Press, 1998); and P. Hall, *Cities of Tomorrow: An Intellectual History of Urban Planning and Design in the Twentieth Century* (Oxford, UK: Blackwell, 2002).

16. A. Scott, "Resurgent Metropolis," 548–64.

17. J. Dodson and N. Sipe, "Oil Vulnerability in the Australian City: Assessing Socio-Economic Risks from Higher Urban Fuel Prices," *Urban Studies* 44 (March 2007): 37–62.

18. R. Heinberg, *The Party's Over: Oil, War, and the Fate of Industrial Societies* (Gabriola Island, BC: New Society Publishers, 2003); C. Campbell, *Oil Crisis* (London: Multi-Science Publishing, 2005); Government Accountability Office, *Crude Oil: Uncertainty about Future Oil Supply Makes It Important to Develop a Strategy for Addressing a Peak and Decline in Oil Production* (Washington, DC: United States Government, 2007); International Energy Agency (IEA), *World Energy Outlook 2008* (Paris: International Energy Agency and Organisation for Economic Cooperation and Development, 2008).

19. Quoted in: G. Monbiot, "When Will the Oil Run Out?" *The Guardian* (London), December 15, 2008.

20. See Kenworthy chap. in this book.

21. Graham and Marvin, *Splintering Urbanism*.

22. J. Dodson and N. Sipe, "Shocking the Suburbs: Urban Location, Housing Debt, and Oil Vulnerability in the Australian City," *Urban Research Program Research Paper 8* (Urban Research Program, Griffith University, Brisbane, 2006); and J. Dodson and N. Sipe, "Shocking the Suburbs: Urban Location, Homeownership, and Oil Vulnerability in the Australian City," *Housing Studies* 23, no. 3 (2008): 377–401.

23. See: Krizek, "Transit Supportive Home Loans."

24. Dodson and Sipe, "Shocking the Suburbs" (2006); and Dodson and Sipe, "Shocking the Suburbs" (2008).

25. Ibid.

26. International Energy Agency, *World Energy Outlook 2008*.

27. P. Marcuse and R. van Kempen, *Globalizing Cities: A New Spatial Order?* (London and Cambridge, UK: Blackwell Publishers, 2000); Walks, "The Social Ecology of the Post-Fordist /Global City?"; and T. Wessel, "Social Polarisation and Socioeconomic Segregation in a Welfare State: The Case of Oslo," *Urban Studies* 37, no. 11 (2000): 1947–67.

28. Graham and Marvin, *Splintering Urbanism*.

3 Full Cost Analysis of Petroleum Consumption[1]

TODD LITMAN

Petroleum production and distribution imposes various economic, social, and environmental costs, including many that are *nonmarket* (involving resources that are not normally traded in competitive markets, such as human health and environmental quality), and *external* (costs are imposed on others).[2] It is important to consider all of these impacts when making policy and planning decisions, such as evaluating energy conservation policies and efficient fuel-tax levels.

This chapter provides a comprehensive review of various external costs resulting from petroleum production, importation, and distribution. It considers four major cost categories: financial subsidies, economic and national security costs of importing petroleum, environmental damages, and human health risks. This chapter does not account for the *internal* costs of petroleum (the direct costs to users) or the external costs that are associated with fuel use, such as the costs of building roadway facilities, vehicle congestion and accident costs, or tailpipe pollution costs, which are explored in other studies.[3]

Evaluating Costs

Cost refers to the loss of scarce and valuable resources, which can include money, land, productivity, human health and life, and natural resources such as clean air and water. What most people call a *problem* economists may call a *cost*, with the implication that its impacts can be quantified (measured).

Costs and benefits (together called *economic impacts*) have a mirror-image relationship: costs can be defined as loss of benefits, and benefits are often measured based on reductions in costs. For example, pollution-reduction benefits are measured based on the resulting reduction in damages to natural resources and human health.

Some impacts are relatively easy to quantify because they involve *market goods*—that is, resources commonly traded in a competitive market. For example, if pollution reduces fishery productivity the costs can be calculated based on fishers' lost income and profits. Impacts involving *nonmarket goods*—that is, resources that are not normally traded in a market, such as the value of enjoying recreational fishing—as well as the broader value of ecological integrity, tend to be more difficult to quantify. Several techniques, discussed below, are used to quantify and monetize nonmarket impacts.[4]

1. *Control or Prevention Costs*
A cost can be estimated based on prevention, control, or mitigation expenses. For example, if industry is required to spend $1,000 per ton to reduce emissions of a pollutant, we can infer that society estimates those emissions to impose costs at least that high. If both damage costs and control costs can be calculated, the lower of the two are generally used for analysis on the assumption that a rational economic actor would choose prevention if it is cheaper, but would accept damages if prevention costs are higher.

2. *Compensation Rates*
Legal judgments and other damage-compensation rates can sometimes help monetize nonmarket costs. For example, if pollution victims are compensated at a certain rate, this can be estimated to represent their damage costs. However, many damages are never compensated. For example, damages can result from many dispersed sources, making fault difficult to assign; damages are often difficult to monetize; ecological systems often lack legal status for compensation; and little compensation may be paid for the deaths of workers who have no dependents. In addition, it is considered poor public policy to provide very generous damage compensation, since this may encourage some people (those who place relatively low value on their injuries) to take excessive risks or even to cause accidents in order to receive compensation. As a result, total environmental and health costs, and society's willingness to prevent such damages, is often much greater than compensation costs.

3. *Hedonic Methods (also called "Revealed Preference")*
Hedonic pricing infers values for nonmarket goods from their effect on market prices, property values, and wages. For example, if houses on streets with heavy traffic are

valued lower than otherwise comparable houses on low-traffic streets, the cost of traffic (conversely, the value of neighborhood quiet, clean air, safety, and privacy) can be estimated. If employees who face a certain discomfort or risk are paid more than otherwise comparable employees who don't, the costs of that discomfort or risk can be estimated.

4. Contingent Valuation (also called "Stated Preference")

Contingent valuation involves asking people how much they value a particular non-market good. For example, residents may be asked about their *willingness to pay* for a particular improvement in environmental quality or safety, or their *willingness to accept* compensation for a particular reduction in environmental quality or safety. Although the analysis methodologies are the same, the results often differ. For example, people may only be willing to pay a $20 per month rent premium for a 20 percent reduction in noise impacts (perhaps by moving to a quieter street or installing sound insulation in their homes), but would demand $100 per month in compensation for a 20 percent increase in residential noise, due to a combination of budget constraints (an inability to pay more rent) and consumer inertia (the tendency of people to become accustomed to a particular situation and so to place a relatively small value on improvements and a relatively large value on degradation). Which perspective is appropriate depends on *property right*—that is, people's right to impose impacts on others. If safety and environmental quality are considered rights, then traffic-crash risk and pollution-emission costs should be based on recipients' willingness to accept incremental harms. If people are considered to have a certain right to impose risk or release pollution, then crash and pollution costs should be calculated based on victims' willingness to pay for an incremental reduction in risk and environmental degradation.

5. Travel Cost

This method uses visitors' travel costs (monetary expenses and time) to measure consumer surplus provided by a recreation site such as a park or other public lands.

Many published cost estimates only reflect a portion of total damages.[5] For example, some pollution cost estimates reflect only direct impacts on a particular industry, or severe health impacts (those that require medical treatment or cause disability and death). Other losses, such as impacts on recreation activity, less-severe illnesses, and ecological integrity, are often excluded. It is important that people working with such values understand the scope and assumptions used in analysis. When reporting costs from a particular study, it is important to define which costs are included, and identify any possible costs that are excluded. For example, when reporting estimated air pollution costs, it would be most accurate to say that ozone and particulate costs average 5¢ per vehicle-mile than to say that air pollution costs average 5¢ per vehicle-mile.

As much as possible, cost estimates should be based on *lifecycle impact analysis* (LIA), which includes costs incurred during production, distribution, use, and disposal.[6] Energy used in production and distribution is sometimes called *embodied energy*. Embodied energy typically represents 25–50 percent of total transportation energy use, depending on mode, as illustrated in figure 3.1.

Petroleum Production, Consumption, and Spill Trends

Several trends will affect the magnitude of future external petroleum costs:

1. Production[7]

Production of conventional, land-based petroleum is currently declining in the United States and is expected to start declining worldwide in the next few years, a trend often called *peak oil*.[8] Total US production is predicted to increase during the next three

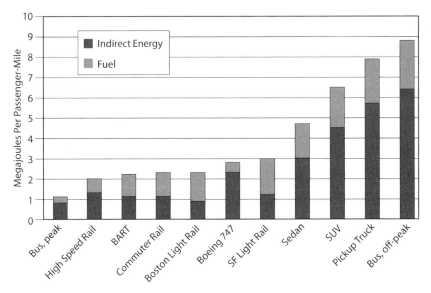

Figure 3.1   Life-cycle energy consumption and emissions: this figure compares fuel and indirect energy (energy used in vehicle and facility construction, as well as maintenance) for various transport modes. (Source: Mikhail V. Chester and Arpad Horvath, "Environmental Assessment of Passenger Transportation Should Include Infrastructure and Supply Chains," *Environmental Research Letters* 4 [2009]

decades due to enhanced oil recovery techniques, increased offshore oil production and increased production of unconventional fuels (such as biofuels and liquefied coal), but these are expensive and speculative.[9] Petroleum will not suddenly run out, but is expected to become more expensive due to rising production costs and increasing international demand. For example, the US Energy Information Administration's *International Energy Outlook 2011* "Reference Case" (most-likely scenario) predicts that real (inflation-adjusted) oil prices will be $95 per barrel in 2015 and increase slowly to $125 per barrel in 2035, but it is possible that prices will increase more rapidly. (At this writing in October 2012, international oil prices are already fluctuating around the $95 per barrel projected price.)

Higher oil prices are likely to increase the production of alternative fuels including offshore oil, tar sands, oil shales, liquified coal, and biofuels such as ethanol and biodiesel. These also have significant external costs, as summarized in table 3.1.

Table 3.1 Alternative transport fuels compared with conventional petroleum[a]

Fuel Type	Benefits	Costs
Offshore oil wells	Can increase domestic production	Additional environmental damages and risks associated with surveying and developing wells, as well as producing and transporting oil.
Biofuels (vegetable oils and ethanol)	Renewable; biodegradable; domestically produced; may reduce some air pollutants	Increases food costs; increases agricultural pollution (such as nitrogen loading of groundwater); nonrenewable fossil fuels are used in production; tends to reduce fuel economy.
Natural gas	Can increase domestic production; reduced air pollutants	Externalities of gas production and transport; nonrenewable fossil fuel source; driving range is generally reduced; limited availability; extra tank is often required, which reduces cargo space
Electricity	Zero tailpipe emissions; widely available	Externalities from electricity production; additional vehicle and battery costs; limited range and performance
Synthetic fuels (tar sands, oil shales, liquefied coal)	Abundant supply exists	Significant environmental damages from extraction and processing; high carbon emissions (10–20% higher per unit of energy than petroleum); high production costs

[a] Consumer Reports. *Alternative Fuels: How They Compare* (Greener Choices, 2006), www.greenerchoices.org/products.cfm?product=alternat&pcat=autos.

2. *Consumption*[10]

World oil consumption is currently about 86 million barrels per day (MBPD), and this is projected to increase to 112 MBPD in 2035.[11] The United States currently consumes about 18.7 MBPD, or 6.8 billion barrels annually.[12] U.S. residents consume about twice as much petroleum per capita as residents of other wealthy countries.[13] However, petroleum consumption has been flat or has declined in the United States and many other mature developed countries, while consumption is increasing rapidly in developing countries such as India and China, so the US share of world petroleum consumption has declined from 46 percent in 1960 to 22 percent in 2009. This trend is expected to continue into the future.[14] The United States currently imports about half the petroleum it consumes; this share is projected to decline in future years if domestic production increase as projected, but even optimistic scenarios predict that the United States will continue to import a major portion of its liquid fuels.[15]

3. *Oil Spills*

Petroleum production, processing, and distribution can result in oil spills that range from small to large. In response to regulations, liability costs, and public-image concerns, the oil industry (including shippers and distributors) has worked to reduce spills and their damages. The frequency and total volume of oil spills declined between 1970 and 2000, particularly by oil tankers, due to improved prevention.[16] However, there are still numerous major oil spills (more than 1,000 tonnes) every year, and catastrophic spills (more than 50,000 tonnes) at least once a decade, as shown in table 3.2. This indicates that, despite efforts to minimize accidents, major oil spills continue to occur.

For every major oil spill there are probably dozens or hundreds of smaller spills, including leaking storage tanks and careless disposal of waste oil by mechanics. In addition, some new petroleum-production techniques introduce new water pollution threats. For example, some enhanced oil-recovery techniques produce large quantities of brine, which may contain salts and various toxic and radioactive substances. Tar sands and oil shale processing often releases toxic chemicals into surface and groundwater during the separation process and through the drainage of rivers, and into the air due to the release of carbon dioxide and other emissions.

Major oil spills occur regularly, despite prevention efforts. This suggests that oil spills and water pollution are, to some degree, an unavoidable result of the production and distribution of petroleum and related products. Although oil spill prevention and cleanup technologies continue to improve, some risks are likely to increase, including those associated with offshore and Arctic area spills, and releases of pollutants into ground and surface water during the production of alternative fuels.

Table 3.2 Selected examples of major oil spills[a]

Name	Location and Date	Estimated Volume (tonnes)
Peace River Rainbow pipeline spill	Alberta, Canada, April 2011	3,800
Talmadge Creek oil spill	Calhoun, Michigan, July 2010	2,800–3,250
MT Bunga Kelana 3	Singapore, Singapore Strait, May 2010	2,000–2,500
2010 ExxonMobil oil spill	Nigeria, Niger Delta, May 2010	3,246–95,500
Deepwater Horizon	Gulf of Mexico, April–July 2010	492,000–627,000
Montara oil spill	Australia, Timor Sea, August 2009	4,000–30,000
2008 New Orleans oil spill	New Orleans, Louisiana, July 2008	8,800
2007 Statfjord oil spill	Norwegian Sea, December 2007	4,000
Korea oil spill	South Korea, Yellow Sea, December 2007	10,800
Jiyeh power station oil spill	Lebanon, July 2006	20,000–30,000
Bass Enterprises	Cox Bay, Louisiana, August 2005	12,000
Tasman Spirit	Pakistan, Karachi, July 2003	28,000
Erika	France, Bay of Biscay, December 1999	15,000–25,000
Sea Empress	United Kingdom, Pembrokeshire	40,000–72,000
MV Braer	United Kingdom, Shetland, January 1995	85,000
Aegean Sea	Spain, A Coruña, December 1992	74,000
Fergana Valley	Uzbekistan, March 1992	285,000
ABT Summer	Angola, May 1991	260,000
MT Haven	Mediterranean Sea, April 1991	144,000
Khark 5	Las Palmas de Gran Canaria, December 1989	70,000
Exxon Valdez	Prince William Sound, Alaska, March 1989	37,000–104,000
Odyssey	Nova Scotia, November 1988	132,000

[a] Wikipedia, *List of Recent Oil Spills*, en.wikipedia.org/wiki/List_of_oil_spills

External Cost Categories

There are various categories of external petroleum costs, including production subsidies, economic and national security costs of importing oil, and environmental and human health damages from petroleum production and distribution. This chapter discusses each of these categories.

1. *Financial and Economic Subsidies*

Energy industries benefit from various financial subsidies and tax exemptions.[17] These include accelerated depreciation of energy-related capital assets, under-accrual for oil- and gas-well reclamation, low royalties for extracting resources from public lands, public funding of industry research and development programs, and subsidized water infrastructure for oil industries.[18] Koplow and Dernbach identify the following major energy subsidies:[19]

- Defending Persian Gulf oil shipping lanes
- Subsidizing water infrastructure for coal- and oil-industry use
- Federal spending on energy research and development
- Accelerated depreciation of energy-related capital assets
- Under-accrual for reclamation and remediation at coal mines and oil and gas wells
- The ethanol exemption from the excise fuel tax.

By considering approximately 75 programs and tax breaks, Koplow estimates that US federal energy-sector subsidies totaled $49 to $100 billion annually in 2006, of which about half are for petroleum ($25 to $50 billion), indicating that petroleum subsidies average about $3.50 to $7.00 per barrel, and significantly more if state-level subsidies are also included.[20] Another study estimates that US fossil-fuel subsidies, including obscure tax-code provisions such as the Foreign Tax Credit (which allows royalty payments to foreign governments to be considered as corporate income taxes) and the Credit for Production of Nonconventional Fuels (which provides a tax credit for the production of certain fuels including oil shales, tar sands and coal-based synthetic fuels) total approximately $10 billion annually, or about $1.50 per barrel.[21]

Other countries also provide large energy production subsidies. Metschies identifies approximately 40 countries where gasoline and fuel retail prices are below international gasoline prices, indicating significant subsidy.[22] International Monetary Fund analysis estimated that in 2010 global petroleum-product subsidies totaled almost $250 billion, and $740 billion including tax subsidies, or approximately 1 percent of global GDP.[23] The International Energy Agency estimates that energy subsidies (mostly for oil, gas, and coal) totaled $557 billion, and that eliminating energy subsidies would cut global GHG emissions 10 percent by 2050.[24]

2. *Economic and National Security Costs of Petroleum Importation*
Dependence on imported petroleum imposes macroeconomic costs by reducing economic productivity, employment, and incomes. This cost is indicated by the fact that major oil-price spikes are often followed by economic recessions. Because North America consumes a major share of world petroleum production, high US demand increases international oil prices, which is called a *pecuniary cost of oil use*.[25] This imposes financial costs on oil consumers and increases the wealth transfer from oil consumers to producers, exacerbating other economic costs. These are primarily economic transfers from oil consumers to producers, and so are not necessarily costs from a global perspective, but to the degree that they lead to recessions and reduce international productivity, they can impose international costs.

Petroleum and motor-vehicle imports are major contributors to the US trade deficit. In 2009 the US had a $381 billion trade deficit of which $253 billion was from oil imports and $160 billion from vehicle and vehicle-part imports, offset by $81 billion in vehicle exports, for a $332 net import burden, representing 87 percent of that year's trade deficit.[26] A major federal study estimated that oil dependence cost the US economy $150–250 billion in 2005 when petroleum prices were just $35–45 per barrel,[27] which suggests that, due to higher international oil prices, these costs now total $300–$500 billion annually, equivalent to $85–140 per barrel of imported oil or $44–74 per total barrels of oil consumed in the US. These costs are relatively evenly divided between transfer of wealth from the United States to oil-producing countries, the loss of economic potential due to oil prices elevated above competitive market levels, and disruption costs caused by sudden and large oil-price changes. These estimates do not include military, strategic, or political costs associated with US and world dependence on oil imports. A 2007 federal report estimates the external economic costs of importing oil to the US (defined as "the quantifiable per-barrel economic costs that the US could avoid by a small-to-moderate reduction in oil imports"), excluding military expenditures, totaled $13.60 per barrel (2004 dollars), with a range of $6.70 to $23.25, or about $54 billion annually for the US.[28]

Empirical evidence indicates that, all else being equal, low fuel prices reduce economic productivity, particularly in oil-consuming countries and regions (areas where a significant portion of petroleum is imported), as indicated in figure 3.2. This occurs because low fuel prices encourage increased per-capita fuel consumption, and therefore petroleum importation costs, and tends to create automobile-dependent transport systems. This reduces regional employment and business activity, and it increases total transportation costs, including traffic congestion, infrastructure costs, accidents, and pollution damages.

Described differently, public policies that encourage energy conservation, such as high fuel taxes, tend to support economic development by reducing the economic burden of importing petroleum and reducing total transportation costs. This is true even in oil-producing regions. For example, although Norway is a major petroleum producer it maintains high fuel prices and energy conservation policies, which leaves more oil to export. As a result, Norway has one of the world's highest incomes, a competitive and expanding economy, a positive trade balance, and the world's largest legacy fund (an investment fund for future generations). Other oil producers, such as Saudi Arabia, Venezuela, and Iran, experience relatively less economic development due to low fuel prices that encourage inefficient fuel consumption and increased associated costs such as traffic congestion, accidents, and pollution emissions.

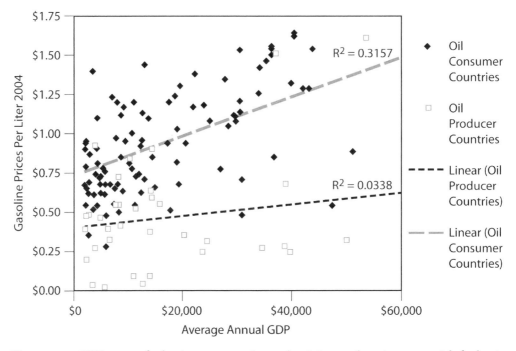

Figure 3.2   GDP versus fuel prices: economic productivity tends to increase with fuel prices, particularly in oil-consuming countries. (Source: Todd Litman, "Evaluating Transportation Economic Development Impacts" [VTPI (www.vtpi.org), 2010], www.vtpi.org/econ_dev.pdf.)

Dependence on imported resources imposes military, political, and economic costs associated with protecting access to foreign petroleum supplies. For example, Persian Gulf military expenditures currently average about $500 billion annually,[29] plus indirect and long-term costs, such as lost productivity and future disability costs from military casualties, uncompensated losses to civilians, and environmental damages.[30] Delucchi and Murphy estimate that 60 percent of Persian Gulf military costs are to maintain access to oil, representing about $300 billion annually.[31] These costs average at least $140 per imported barrel or $74 per total barrel consumed, and possibly significantly more.

Stern estimates that US Middle East military intervention costs, intended to maintain US access to petroleum resources, average about $500 billion annually.[32] He concludes that these military costs come in addition to economic costs of a comparable magnitude, implying that US oil-dependence costs total about $1 trillion annually. The National Defense Council Foundation estimates that the external costs of US oil

Table 3.3 External costs of US oil imports, 2003 and 2006[a]

	2003	2006
Oil-related defense expenditures	$49.1 billion	$137.8 billion
Loss of current economic activity due to capital outflow	$36.7 billion	$117.4 billion
Loss of domestic investment	$123.2 billion	$394.2 billion
Loss of government revenues	$13.4 billion	$42.9 billion
Cost of periodic oil-supply disruptions	$82.5 billion	$132.8 billion
Total	$304.9 billion	$825.1 billion
Job losses	828,400	2,241,000

[a] National Defense Council Foundation, *Hidden Cost of Oil: An Update* (NDCF [www.ndcf.org], 2007), ndcf.dyndns
.org/ndcf/energy/NDCF_Hidden_Cost_2006_summary_paper.pdf.

imports increased from $305 billion in 2003 to $825 billion in 2006, as summarized in table 3.3.

There is debate concerning the portion of military costs that should be charged to petroleum consumers.[33] *Marginal analysis* (reflecting incremental changes in costs from incremental changes in consumption) tends to allocate relatively small costs to consumers since there may be other justifications for overseas military interventions (such as controlling terrorism and establishing democracy), and many military and political costs can be considered fixed in the short- and medium-term, so it is difficult to determine how these costs would decline with reduced fuel consumption.[34] *Cost-recovery analysis* (total costs are charged to users) allocates a larger share of national-security costs to petroleum consumers.

For evaluating policies that specifically affect the amount of petroleum that will be imported (such as the imposition of import duties), these costs should apply specifically to imported oil. For evaluating policies that affect total national fuel consumption (such as general fuel taxes or fuel-efficiency mandates), these costs should apply to total fuel consumed, since the marginal barrel of oil is imported.

3. *Environmental Damages*

Resource exploration, extraction, processing, and distribution cause environmental damages, including habitat disruption from exploration and drilling activity, shorelines spoiled by refineries, noise and water pollution, air pollution such as sour gas (hydrogen sulfide) and greenhouse-gas emissions (see chap. 1), and oil spills. Although newer policies and practices are intended to reduce these impacts, and some damages are compensated, there are significant residual damages, and many impacts are projected to increase with increased development of deep ocean wells and alternative fuels such as tar sands and oil shale.

Pollution emissions that occur during fuel production (as opposed to use) are called *upstream* emissions, which are said to be *embodied* into the final product.[35] According to detailed lifecycle analysis, embodied energy and emissions add about 16 percent to the energy and greenhouse emissions that occur during fuel use.[36]

Analysis of various US oil spills indicates that cleanup and damage-compensation costs range from less than $300 per barrel ($7 per gallon) for the 1979 *Ixtoc I* spill in the Gulf of Mexico, up to more than $25,000 per barrel ($630 per gallon) for the 1980 *Exxon Valdez* spill in Alaska, with an average of approximately $672 per barrel ($16 per gallon).[37] To the degree that these damages are compensated, they are borne by the oil industry and passed on to consumers. However, many damages are never compensated because they are difficult to quantify, involve ecological services that lack legal status, or are limited by liability caps in state and federal laws.[38] According to surveys, the lower-bound estimate of the public's willingness to pay to avoid the *Valdez* spill's wildlife damages was $2.8 billion, compared with approximately $1.0 billion in total wildlife cleanup and compensation costs.[39] This suggests that total damage costs, and society's willingness to pay to avoid damages, are significantly (perhaps two to five times) higher than the financial costs borne by the oil industry.[40]

As an example, the 2010 *Deepwater Horizon* oil-spill cleanup and compensation costs are predicted to total $20–40 billion.[41] Assuming that one such catastrophic spill occurs each decade, this averages $2–4 billion a year. However, this only includes direct, legally recognized damages from major spills; it excludes smaller spills, "normal" environmental damages caused by petroleum production and processing (oil wells, refineries, and transport facilities), and uncompensated ecological costs, such as losses of existence value, as well as aesthetic value, from destruction of wildlife and landscapes. Production of alternative fuels such as oil sands and liquefied coal is generally considered more environmentally damaging than conventional oil production; it causes landscape damage, consumes large amounts of fresh water, and produces more climate-change emissions per unit of fuel.[42] Some damages, such as irreversible habitat destruction, can have very high costs but lack legal standing.

In addition to current losses, some economists argue that depleting nonrenewable resources deprives future generations of important benefits, implying a moral obligation to conserve resources for the sake of intergenerational equity.[43]

This suggests that the total environmental costs of petroleum production, processing, and distribution are probably many times larger than just current cleanup and compensation costs, perhaps $10–30 billion annually in the United States. This averages $1.50–4.50 per barrel, or 3.8–11.4¢ per gallon of petroleum products consumed.

4. *Human Health Risks*

Resource exploration, extraction, processing, and distribution cause various health risks to people, including processing and distribution accident injuries as well as pollution-related illnesses. In 2006 petroleum production workers had 20.8 fatalities per 100,000 workers, which is much higher than typical service-industry jobs but lower than other heavy industries such as truck drivers (27.5 deaths), coal miners (49.5 deaths), and loggers (87.4).[44] In addition, oil wells and petroleum refineries sometimes emit harmful air and water pollution that may endanger people nearby, leading some areas to be considered "cancer alleys," although the actual magnitude of such risks is difficult to determine.[45]

These costs are partly internalized through worker compensation and liability claims, but, as discussed previously, it is impossible to fully compensate some losses, because, from an individual's perspective, no amount of money can fully compensate for death or severe disability, and it is considered poor public policy to provide overly generous damage compensation because doing so may encourage some people to take excessive risks (for example, workers may be less cautious if they believe that even minor injuries will be generously compensated). These human-health pollution risks are often included in "environmental cost" categories, so it is important to avoid double-counting when calculating monetized cost estimates.

Conclusions

Petroleum production, importation, and distribution can impose a number of external costs. These are costs that people ultimately bear through higher taxes, reduced productivity, environmental damages, and health problems, but are widely dispersed rather than charged directly to consumers based on the amount of petroleum they consume and therefore their contribution to these costs. These external costs tend to be inefficient because they encourage people to consume more petroleum, and therefore impose more total costs, than would occur if consumers bore these costs directly, and they are inequitable because they result in one individual or group imposing costs on others.

Table 3.4 summarizes the various estimates of external costs described in this chapter. It indicates that US external costs of petroleum production, importation, and distribution probably total $635–1,080 billion annually, depending on assumptions, which averages $93–160 per barrel or $2.21–3.78 per gallon. This analysis suggests that for every dollar that consumers spend on petroleum (internal costs), their petroleum consumption imposes $0.63–1.08 in external costs (assuming $3.50 per gallon average prices).

Many published estimates of petroleum external costs consider only a portion of these impacts, and so underestimate total costs and the total benefits of energy conservation. Some of these costs are likely to increase in the future with increased exploitation of higher risk alternative fuels, such as offshore oil, tar sands, and liquefied coal.

This analysis only accounts for the external costs of petroleum production, importation, and distribution. It excludes the environmental costs of petroleum consumption (such as air pollution and climate-change emissions), and the external costs of vehicle use powered by petroleum products (such as road and parking-facility costs, traffic congestion, and accidents).[46]

Implications for Optimal Fuel Policy

This analysis indicates that petroleum production, importation, and distribution impose significant external costs. Although cost estimates vary depending on perspective and assumptions, even lower-bound values indicate that petroleum is significantly underpriced. Vehicle fuel prices would have to increase by half or two-thirds if production subsidies and favorable tax policies were eliminated, if consumers paid directly for the economic and security costs of producing and importing petroleum, and if all environmental and human health costs were fully compensated. This does not include additional external costs of fuel use, such as greenhouse-gas emissions, nor the external costs of vehicle use, such as traffic congestion, parking subsidies, and uncompensated accident damages.

Fuel underpricing may have been justified in the past when petroleum, motor vehicle, and roadway systems were first growing and so were beginning to experience economies of scale (unit costs declined as total consumption increased), but these industries are now mature, and fuel consumption and motor vehicle travel impose significant external costs.

Advocates of underpricing often argue that low fuel prices benefit poor people, but the vast majority of these benefits go to non-poor people who tend to consume the majority of petroleum products.[47] Fuel taxes tend to be regressive (they represent a larger share of budgets for lower- than higher-income households), but this

Table 3.4 Summary of petroleum external costs

Name	Description	Estimates (annual billion dollars)
Production subsidies	Direct government subsidies and tax reductions for petroleum production	$25–50 billion
Economic costs	Economic costs of importing petroleum	$300–500 billion
Security costs	Military expenditures and other costs of maintaining access to foreign oil supplies	$300–500 billion
Environmental costs	Uncompensated environmental damages, including ground and surface water pollution, lost productivity, plus ecological and aesthetic degredation	$10–30 billion
Human health damages	Uncompensated injury and illness costs to workers and nearby residents resulting from petroleum production, distribution	???
Totals	Total external costs	$635–1,080 billion
Total per barrel		$93–160
Total per gallon		$2.21–3.78

Note: Petroleum production, importation, and distribution impose various external costs. These estimates do not include the costs of petroleum consumption (such as pollution and climate-change emissions) nor the external costs resulting from vehicle use (such as road and parking-facility costs, congestion, and accidents), which are often internalized with fuel taxes.

regressivity ultimately depends on the quality of transport options available and how revenues are used.[48] If fuel taxes are used to reduce other regressive taxes, finance new services valued by low-income households (such as walking, cycling, and transit service improvements, or better education and health care services), or are returned as cash rebates, then equity impacts can be neutral or progressive overall.[49]

This indicates that higher fuel taxes and other energy conservation strategies can support economic development and help create more equitable transport systems if implemented gradually and predictably, in conjunction with policies that increase transport-system efficiency and diversity, such as improved walking, cycling, and public transit service, as well as more accessible land-use development.[50]

Notes

1. This chapter summarizes "Resource Consumption External Costs" in: Todd Litman, *Transportation Cost and Benefit Analysis—Techniques, Estimates, and Implications* (publication of Victoria Transport Policy Institute, 2009), www.vtpi.org/tca/tca0512.pdf.

2. European Commission, *ExternE: Externalities of Energy—Methodology 2005 Update* (EC publication, 2005), www.externe.info/brussels/methup05a.pdf; see also: www.externe.info.

3. Litman, *Transportation Cost and Benefit Analysis*; see also: Huib van Essen et al., *Marginal Costs of Infrastructure Use—Towards a Simplified Approach* (publication of CE Delft [www .ce.nl], 2004), www.ce.nl/publicatie/marginal_costs_of_infrastructure_use_%96_towards_a _simplified_approach/456.

4. Transportation Research Board, "Monetary Valuation of Hard-to-Quantify Transportation Impacts: Valuing Environmental, Health/Safety & Economic Development Impacts" (report NCHRP 8–36–61, Economic Development Research Group, 2007), www.statewide planning.org/_resources/63_NCHRP8–36–61.pdf; www.trb.org/nchrp; see also: M. Maibach et al., *Handbook on Estimation of External Cost in the Transport Sector* (publication of CE Delft [www.ce.nl], 2008), ec.europa.eu/transport/costs/handbook/doc/2008_01_15_handbook _external_cost_en.pdf.

5. National Research Council, *Hidden Costs of Energy: Unpriced Consequences of Energy Production and Use* (publication of NRC, 2009), www.nap.edu/catalog/12794.html.

6. Mikhail Chester and Arpad Horvath, "Environmental Life-Cycle Assessment of Passenger Transportation" (report vwp–2008–2, UC Berkeley Center for Future Urban Transport, 2008), www.its.berkeley.edu/volvocenter; see also: www.sustainable-transportation.com; see also: Argonne National Lab, *Greenhouse Gases, Regulated Emissions, and Energy Use in Transportation (GREET) Model*, (ANL publication, 2008), www.transportation.anl.gov/modeling_simula tion/GREET/index.html.

7. Energy Information Administration, *International Energy Outlook 2011* (publication of the US Energy Information Administration [www.eia.gov], 2011), www.eia.gov/forecasts/ieo /pdf/0484(2011).pdf.

8. Robert L. Hirsch, Roger Bezdek, and Robert Wendling, "Peaking of World Oil Production: Impacts, Mitigation, & Risk Management" (report, US Department of Energy [www .netl.doe.gov], 2005), www.netl.doe.gov/publications/others/pdf/Oil_Peaking_NETL.pdf.

9. EIA, *International Energy Outlook 2011*, 27.

10. Oak Ridge National Laboratories, *Transportation Energy Book*, 29th ed. (ORNL publication, US Department of Energy [www.doe.gov], 2010), cta.ornl.gov/data/index.shtml.

11. EIA, *International Energy Outlook 2011*, 1–2.

12. ORNL, *Transportation Energy Book*, table 1.12.

13. NationMaster, "Oil Consumption by Country," *Energy Statistics* (NM [www.nationmas ter.com] report, 2011), www.nationmaster.com/graph/ene_oil_con-energy-oil-consumption.

14. ORNL, *Transportation Energy Book*, table 1.4.

15. ORNL, *Transportation Energy Book*, table 1.7.

16. Dagmar Schmidt Etkin, "Analysis of Oil Spill Trends in the United States and Worldwide" (report presented for Environmental Research Consulting [www.environmental-re search.com] at the International Oil Spill Conference, 2001), www.environmental-research .com/publications/pdf/spill_statistics/paper4.pdf.

17. Global Subsidies Initiative, "Measuring Subsidies to Fossil-Fuel Producers" (report prepared for GSI [www.globalsubsidies.org], 2010), www.globalsubsidies.org/en/research

/gsi-policy-brief-a-how-guide-measuring-subsidies-fossil-fuel-producers; see also: GSI, "Subsidy Watch," www.globalsubsidies.org/en/subsidy-watch; see also: International Institute for Sustainable Development, "Mapping the Characteristics of Producer Subsidies: A Review of Pilot Country Studies" (report, IISD [www.iisd.org], 2010), www.iisd.org/publications/pub .aspx?pno=1327.

18. David Coady et al., "Petroleum Product Subsidies: Costly, Inequitable, and Rising" (report, International Monetary Fund [www.imf.org], 2010), www.imf.org/external/pubs/ft /spn/2010/spn1005.pdf.

19. Doug Koplow and John Dernbach, "Federal Fossil Fuel Subsidies and Greenhouse Gas Emissions: A Case Study of Increasing Transparency for Fiscal Policy," *Annual Review of Energy and the Environment* 26 (2001): 361–89, arjournals.annualreviews.org/loi/energy; see also: www.mindfully.org/Energy/Fossil-Fuel-Subsidies.htm.

20. Doug Koplow, *Subsidy Reform and Sustainable Development: Political Economy Aspects* (publication of the Organisation for Economic Co-operation and Development [OECD, www .oecd.org], 2007), www.earthtrack.net/earthtrack/library/SubsidyReformOptions.pdf.

21. Environmental Law Institute, "Estimating US Government Subsidies to Energy Sources: 2002–2008" (report, ELI [www.eli.org], 2009), www.elistore.org/Data/products /d19_07.pdf.

22. Gerhard P. Metschies, "Fuel Prices and Taxation: With Comparative Tables for 160 Countries" (report prepared for GTZ [www.gtz.de/en], 2005), www.internationalfuelprices .com.

23. Coady et al., "Petroleum Product Subsidies"; see also: International Energy Agency, *Analysis of the Scope of Energy Subsidies and Suggestions for the G-20 Initiative* (joint report of IEA, OPEC, OECD, and World Bank, 2010), www.oecd.org/dataoecd/55/5/45575666.pdf.

24. Organization for Economic Cooperation and Development, "Global Warming: Ending Fuel Subsidies Could Cut Greenhouse Gas Emissions 10%, Says OECD" (report, OECD [www .oecd.org], 2010), www.oecd.org/document/30/0,3343,en_2649_34487_45411294_1_1_1_1,00 .html.

25. Mark Delucchi, "The Social-Cost Calculator (SCC): Documentation of Methods and Data, and Case Study of Sacramento" (report UCD-ITS-RR-05–37, Institute of Transportation Studies [www.its.ucdavis.edu], 2005), www.its.ucdavis.edu/publications/2005/UCD-ITS -RR-05–18.pdf.

26. Kimberly Amadeo, *The US Trade Deficit* (About.com Guide, 2010), useconomy.about .com/od/tradepolicy/p/Trade_Deficit.htm.

27. David Greene and Sanjana Ahmad, *The Costs of Oil Dependence: A 2005 Update* (publication of US DOE [www.doe.gov], 2005), cta.ornl.gov/cta/Publications/Reports/ORNL _TM2005_45.pdf.

28. Paul N. Leiby, *Estimating the Energy Security Benefits of Reduced US Oil Imports* (publication of Oak Ridge National Laboratory [www.ornl.gov], 2007), www.epa.gov/oms/renewable fuels/ornl-tm–2007-028.pdf.

29. Roger J. Stern, "United States Cost of Military Force Projection in the Persian Gulf, 1976–2007," *Energy Policy* 38 (2010): 2816–25.

30. David L. Greene, "Measuring Energy Security: Can the United States Achieve Oil Independence?" *Energy Policy* 38, no. 4 (April 2010): 1614–21.

31. Mark Delucchi and James Murphy, "US Military Expenditures to Protect the Use of Persian-Gulf Oil for Motor Vehicles," *Energy Policy* 36 (2008).

32. Roger J. Stern, "United States Cost of Military Force Projection in the Persian Gulf, 1976–2007," *Energy Policy* 38 (2010): 2816–25, www.princeton.edu/oeme/articles/US-miiltary -cost-of-Persian-Gulf-force-projection.pdf.

33. Delucchi and Murphy, "US Military Expenditures."

34. Ibid., 2253–64.

35. Chester and Horvath, "Environmental Life-Cycle Assessment."

36. Mikhail V. Chester, "Life-Cycle Environmental Inventory of Passenger Transportation in the United States," in *Environmental Life-Cycle Assessment of Passenger Transportation: An Evaluation of Automobiles, Buses, Trains, Aircraft, and High Speed Rail* (publication of the Institute of Transportation Studies [www.sustainable-transportation.com], University of California, Berkeley, 2008), http://escholarship.org/uc/item/7n29n303. Gasoline-anddiesel-fuel- -embodied energy is described on pages 52–53.

37. Mark A. Cohen, "Taxonomy of Oil Spill Costs—What Are the Likely Costs of the *Deepwater Horizon* Spill?" (report, Resources for the Future [www.rff.org], 2010), www.rff.org/rff /documents/RFF-BCK-Cohen-DHCosts_update.pdf.

38. Nathan Richardson, "*Deepwater Horizon* and the Patchwork of Oil Spill Liability Law" (report, Resources for the Future [www.rff.org], 2010), www.rff.org/RFF/Documents/RFF -BCK-Richardson-OilLiability_update.pdf.

39. Cohen, "Taxonomy of Oil Spill Costs."

40. Mark A. Cohen, *Deterring Oil Spills: Who Should Pay and How Much?* (publication of Resources for the Future [www.rff.org], 2010), www.rff.org/Publications/Pages/PublicationDe tails.aspx?PublicationID=21161.

41. Jonathan L. Ramseur, "Liability and Compensation Issues Raised by the 2010 Gulf Oil Spill" (report, Congressional Research Service [www.loc.gov/crsinfo], 2011), assets.opencrs .com/rpts/R41679_20110311.pdf.

42. Canadian Assocation of Petroleum Producers, *Environmental Challenges and Progress in Canada's Oil Sands* (publication of CAPP [www.capp.ca], 2008), www.capp.ca/getdoc .aspx?DocID=135721.

43. J. Gowdy and S. O'Hara, *Economic Theory for Environmentalists* (Boca Raton, FL: St. Lucie Press [www.crcpress.com], 1995).

44. Bureau of Labor Statistics, "Fatal Occupational Injuries, Employment, and Rates of Fatal Occupational Injuries, 2006" (report, BLS [www.bls.gov], US Department of Labor, 2007), www.bls.gov/iif/oshwc/cfoi/CFOI_Rates_2006.pdf.

45. WBK Associates, "Health Implications of Petroleum Refinery Air Emissions" (report, Canadian Council of Ministers of Environment [www.ccme.ca], 2003), www.ccme.ca/assets /pdf/wbk_ex_summ.pdf.

46. Litman, *Transportation Cost and Benefit Analysis*; Maibach et al., *Handbook on Estimation of External Cost*.

47. Javier Arze del Granado and David Coady, "The Unequal Benefits of Fuel Subsidies: A Review of Evidence for Developing Countries" (report, International Monetary Fund [www .imf.org], 2010), www.imf.org/external/pubs/cat/longres.cfm?sk=24184.0.

48. Center on Budget and Policy Priorities, "Climate-Change Policies Can Treat Poor Families Fairly and Be Fiscally Responsible" (report, CBPP [www.cbpp.org/pubs/climate-brochure .htm], 2007).

49. Aaron Golub, "Welfare and Equity Impacts of Gasoline Price Changes under Different Public Transportation," *Journal of Public Transportation* 13, no. 3 (2010): 1–21, www.nctr.usf .edu/jpt/pdf/JPT13–3.pdf.

50. United Nations Environment Programme, *Reforming Energy Subsidies* (publication of UNEP [www.unep.org], 2008), www.unep.org/pdf/pressreleases/reforming_energy_subsi dies.pdf.

4 How Does Induced Travel Affect Sustainable Transportation Policy?

ROBERT B. NOLAND AND CHRISTOPHER S. HANSON

Induced travel has long been debated among transportation professionals and frequently ignored by planners when considering policy. (Induced travel refers to the observation that congested roads quickly gain new traffic after they have been expanded.) In fact, this has been observed since Western nations began to motorize and to construct major road facilities.[1] Policy initiatives to implement more sustainable transportation systems frequently overlook the role that increased road capacity can play in undermining the objective of achieving greater sustainability. This chapter examines the theory and available evidence on induced travel effects and links this knowledge to policy considerations for achieving sustainable solutions, including the impact on oil dependence.

Transportation modeling and forecasting systems do not typically account for induced travel effects and lead to systematic overestimation of congestion-reduction benefits. This chapter will cover these issues, beginning with a discussion of the basic theoretical issues of how changes in road capacity affect behavior and some discussion of why this has been controversial, followed by a review of empirical estimates of these effects, with a focus on causal models that can more effectively determine whether a change in road capacity actually *causes* a change in travel behavior.[2] We conclude with a discussion of policy issues for those seeking more sustainable transportation systems.

Underlying Economic Theory and Behavioral Factors

Travel has been described as a classic case of a normal good.[3] An increase in the capacity (i.e., the supply) of the highway system reduces the cost of travel, resulting from the reduction in travel time due to reduced congestion as well as greater access which shortens distances (or makes trips faster).[4] Consumption levels are determined by the supply of opportunities to travel as well as the overall demand for travel, regulated by the cost (primarily time) and individual budget constraints (again, primarily time constraints).[5] While the cost for parking, fuel, and maintenance can also affect travel choices,[6] the literature suggests that highway users are more sensitive to the time a trip takes.[7] The same is true of transit users.[8] It is possible to calculate how much travel demand increases as the price drops. This results in derivation of what is known as an elasticity of demand. The historical practice of transportation planners has been to treat travel demand as unresponsive to price, that is, it is considered perfectly inelastic and would have a vertical demand curve.[9]

Figures 4.1 and 4.2 graphically illustrate these simple relationships. Figure 4.1 shows how an exogenous increase in supply,[10] represented by the downward shift in the supply curve (from S1 to S2), affects travel demand; with the inelastic demand curve, there is no change in demand (Q1) and any observed changes are attributed to exogenous increases in demand (Q3) from, for example, population increases or other economic factors. Much of the confusion over the relative importance of induced travel is derived from the assumption that travel demand is inelastic. Any changes in demand are assumed to occur from exogenous population growth and economic factors; however, these can have a spatial component with major impacts on total demand for travel.

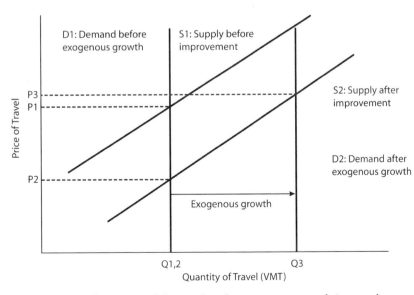

Figure 4.1   Inelastic travel demand and exogenous growth in travel.

Figure 4.2, alternatively, shows a downward-sloping demand curve representing an elastic demand response to changes in cost. In this case, any increase in supply (to S2) corresponds to an increase in demand (Q2). While the costs (i.e., the amount of travel delay) may be less than previously (P2 is less than P1), the reduction is less than that of an inelastic demand response. Exogenous growth still can reduce this benefit and would do so more rapidly (shifting demand to Q3). However, in a full economic modeling system, any exogenous growth may also be caused by the change in travel costs; that is, if we believe that reduced transportation costs increase economic growth, then these factors should be wholly endogenous (or internal to the system).

The time budget literature cited in Noland and Lem suggests that the time that people allocate for travel is fairly stable,[11] and has remained stable over time; for example, Zahavi and Talvitie report that daily travel times across a number of countries are fairly consistent.[12] By relating travel time to speed and therefore distance, Zahavi and Talvitie imply that demand for travel is elastic. However, Zahavi and Ryan report that total travel times are quite consistent over time.[13] As travel times are reduced, travel becomes less expensive to consumers and more travel is consumed, up to a given daily limit. Noland and Lem acknowledge the theoretical possibility that overall reductions in the generalized cost of travel could lead to increased daily time budgets.[14]

Travel-demand elasticities are sensitive to time horizon. Over the short term, travelers change their use of the existing transportation system. An increase in capacity results in an immediate reduction in congestion. That reduction is captured conceptually in Downs's triple convergence of responses to congestion reduction,[15] which is widely cited in the induced travel literature.[16] When travel on a highway becomes faster because an increase in capacity reduces congestion, the first effect is convergence of travel to peak periods, since previously travelers had shifted to off-peak (or shoulder) periods due to congestion. Another short-term reaction includes route shifting, away from parallel routes that are now relatively slower. Finally, Downs included shifts away from slower modes, such as public transit. Noland also noted that in the short term, destinations may change so that trips cover longer distances, and trips may also be made more frequently. All these short-run effects can occur fairly rapidly.[17]

Noland and Cowart note that transportation planners have resisted incorporation of induced travel effects on grounds that travel is a derived demand; that is, it is a by-product of other activities that people engage in.[18] This is consistent with the belief of transportation planners that changes in travel demand are not affected by changes in the cost (or travel time) that individuals face. However, the overall cost of engaging in activities (i.e., those that generate the derived demand) is a critical component of how

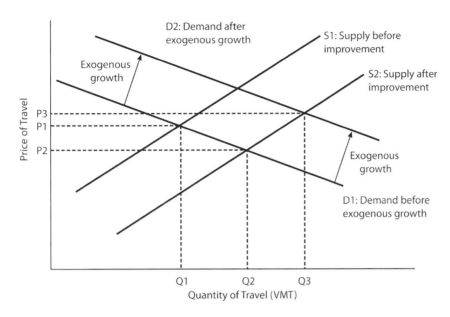

Figure 4.2   Elastic travel demand and exogenous growth in travel.

new capacity affects travel. Increased road capacity can shape the location of those activities that generate demand for travel. Put another way, this is the fundamental way in which land use can change in response to new accessibility patterns and has long been part of basic theories of urban economics.[19]

Urban economics describes the bid-rent function between travel costs (time) and land values.[20] As travel times to access various economic activities decrease, land values increase because consumers make an explicit trade-off between how much they expend on land rent and how much they spend on travel (including time). In simple models, land values are highest near the most accessible locations and are an inverse function of distance from desirable locations.[21] This increase is most notable in places at greater distance from the most desirable locations, which previously had little or no value for development. Thus, one can easily explain how increased road capacity can induce new development at the urban edge and also intensify development in the urban core (assuming that external costs, such as congestion, do not outweigh the benefits of a central location).

These effects can be shown graphically as in figure 4.3. The basic trade-off is displayed by the decline in land value with increasing distance from more accessible locations (typically the center of an urban area). With an increase in road capacity, this line shifts upward and out, increasing the value of all land. The value of land then

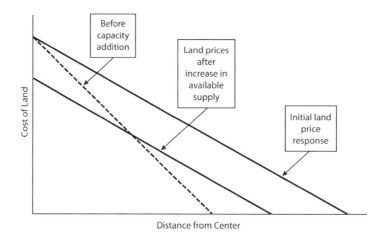

Figure 4.3   Distribution of benefits of accessibility increases.

shifts to the left due to the increased supply of land offsetting the initial increase in value. In simplified terms, this represents, for example, new housing supply leading to a relative reduction in the price of housing. This can also represent increased commercial and retail development of land leading to decreased cost of commercial and retail products, especially if they can take advantage of the increased scale economies offered by the availability of cheaper land. All of these "long-term" effects, attributable to new road capacity, further increase total travel, and thus represent the long-term induced travel impacts.

New road capacity can increase consumer surplus by providing more mobility, whether it reduces congestion in the long term or not. However, this increased consumer surplus is of lesser value than previously existing travel, since prior to the reduction in congestion, it was suppressed.[22] This can be quantified via the "rule of half," which holds that the benefits of induced units of travel are worth half of what units of previously existing travel are worth and that this can be interpreted as a mobility benefit.

An alternative is to consider the change in land accessibility and how this is capitalized into land-value changes. Thus, the beneficiaries are those who currently own the more-accessible land, rather than those who experience increased mobility.[23] A secondary effect is that consumers also capture some of the benefits through increased supply of housing and other products. A further issue in terms of evaluation of economic effects is the potential to encourage or discourage agglomeration (or clustering and concentration) of economic activities. Graham suggests that congestion reduction can improve the external productivity gains achievable from firm agglomeration, but does not distinguish road from public transit effects.[24]

One additional consequence of building increased highway capacity is that it can undermine existing public transit. Public transit can be efficient in terms of energy consumption and emissions. However, it derives its ability to compete with private-vehicle transportation from the frequency with which it can run, and the fares that are charged. The Downs-Thomson paradox addresses the equilibrium between a congested route and public transit.[25] The paradox demonstrates how an increase in road capacity can lead to much worse congestion than was originally present, based on a reduction in public transit service. This can be shown if one considers a train line running parallel to the road in which the operator breaks even by only running trains when they are filled to capacity. As a result the frequency of train runs is a function of total ridership, which determines maximum travel time, including waits. If the parallel congested road has its capacity increased, some travelers will shift away from using the train. As riders shift to the competing roadway the train loses ridership and therefore cuts service (or raises fares). This makes the train less competitive, resulting in further shifts to car usage. Eventually the train is shut down as it is no longer competitive. In the end congestion is potentially worse on the parallel road than before the capacity increase. In fact, price elasticities for transit are fairly small but elasticities for service frequency can be considerably higher,[26] lending credibility to the Downs-Thomson paradox.

Theoretically, we can conclude that one would expect to find that increases in road capacity are likely to increase total travel, especially when projects are aimed at reducing congestion. However, even roads that simply provide greater access under conditions of no congestion may facilitate increased development that, in turn, leads to increased travel. While the theory is straightforward, empirically estimating this effect can be problematic and developing forecasting techniques may not be straightforward. In the next section we review recent research that has examined the empirical evidence for these effects.

Empirical Research Evidence
Empirical estimates of induced travel are well documented in the literature. The majority of these demonstrate that a statistically significant relationship can be found between lane-miles of road capacity and vehicle-miles of travel. These studies typically use aggregate data and multivariate approaches to examine this association. Some go further to examine the endogeneity of traffic growth—that is, whether traffic growth in itself generates the construction of new road facilities. The majority of the empirical

analyses demonstrate that there is a relationship between new roads and extra traffic that is generated; this includes those analyses that controlled for endogenous effects.

The majority of studies in this area have estimated multivariate regression models using area-based aggregations (e.g., state, county, or metro area) of both lane-miles and VMT using cross-sectional time-series approaches. Exogenous factors, such as income, fuel prices, and population, are typically controlled for. Thus, travel is compared on the aggregate level of facilities by region, producing a demand elasticity for travel. Controlling for endogeneity of road construction can be problematic, but has been tackled by several studies.

Regression models typically employ VMT or the logarithm of VMT as the dependent variable. VMT provides a continuous variable that captures changes in vehicular travel including travel from new trips, the added distance from longer trips, and route changes, which are generally recognized as increases in travel.[27] Although these studies use a direct measure of the dependent variable,[28] they do not as directly measure the impact of changes in travel demand on emissions,[29] which are generally linked to the flow characteristics of the traffic network as well as the mix of vehicles in the fleet.

Much of the analysis uses cross-sectional time-series (panel) data on spatial units of analysis (i.e., counties, urban areas, or states). Dummy variables are generally used to capture fixed effects of sub-regions and time periods to control for those factors associated with a given region or that may change over time.

A variety of studies have attempted to tackle these issues, and we present a summary of results in table 4.1. The first detailed analysis was conducted by Hansen et al.[30] and also by Hansen and Huang,[31] who formulated the following functional form, which others have generally followed:

$$\text{Log}(\text{VMT}_{it}) = \alpha_i + \beta_t + \Sigma_k \lambda^k \log(X_{kit}) + \Sigma_{l=0}^{L} \omega^l \log(LM_{it-l}) + \varepsilon_{it}$$

where:

VMT_{it}	represents the VMT in area i in time period t;
α_i	represents the fixed area effects for area i;
β_t	represents the fixed time effects for time period t;
X_{kit}	represents the values of a series of confounding variables k;
LM_{it-l}	represents some measure of a lane-miles increase in region i for lag period t-l;
λ^k, ω^l	represent coefficients for confounding factors and a lagged estimate of the dependent variable, respectively; and
ε_{it}	is an error term.

Table 4.1 Parameter estimates from induced-travel regression models

Reference	Scale	Fixed effects		Causality	Elasticities	
		Area	Time		Short-term	Long-term
Models with aggregate data: all with lane-mile elasticities						
(Hansen et al. 1993)	Facility	X			0.2–0.3	0.3–0.4
(Hansen, Huang 1997)	County	X	X	Lag model	0.21	0.6–0.7
(Hansen, Huang 1997)	Metro	X	X	Lag model	0.19	0.9
(Fulton et al. 2000)	County	X	X	Granger test	0.2–0.6	
(Noland, Cowart 2000)	Metro	X	X	Instrumental variable model	0.28	0.90
(Noland 2001)	States	X	X	Distributed lag model	0.2–0.5	0.7–1.0
(Cervero, Hansen 2002)	County			Simultaneous equations		
VMT dependent	County	X	X	Granger test	0.59	0.79
LM dependent	County	X	X	Granger test	0.33	0.66
(Cervero 2003)						
Direct	Facility	X	X	4-element path model	0.24	0.81
Indirect	Facility	X	X	4-element path model	0.10	0.39
(Duranton, Turner 2009)	States	Cross-sectional		Instrumental variable model	0.92–1.32	
(Hymel, Small, & Van Dender 2010)	States	X	X	3-stage least squares	0.037	0.186
(Rentziou, Gkritza, & Souleyrette 2011)	States	Random effects		Error component model	Urban, 0.256	Rural, 0.068

Models with disaggregate data	Scale	Type of elasticity	Elasticities
Strathman et al. (2000)			
Direct	Corridor	Lane-miles	0.29
Indirect	Corridor	Lane-miles	0.033
Barr (2000)	Corridor	Travel time	−0.3 to −0.5

The use of the logarithmic form has two advantages. First, it minimizes any statistical problems caused by including regions with large variances in size.[32] Second, it allows one to interpret coefficient estimates as elasticities, although this also assumes that elasticities are constant and independent of the existing level of road capacity. In itself this is not an unrealistic assumption, but also, as these studies are mainly concerned with identifying a statistically significant effect, the absolute magnitude of the estimated elasticity is less critical, although this would matter if the model is used for forecasting purposes.

The studies listed in table 4.1 all show statistically significant effects of lane mileage affecting vehicle-miles of travel. There is some variety in the data used and the statistical methods used, and thus we see variation in the actual parameter estimates derived from these models. This is a potential problem for any forecasting methodology that seeks to determine how road capacity affects demand.

As an example of the range of parameter estimates, Hymel, Small, and Van Dender[33] use the same data as Noland.[34] Their modeling approach was different in that it was designed to control for simultaneous effects—that is, how vehicle travel itself may affect the building of new capacity.[35] They find a much smaller elasticity estimate, even when using the shorter time series of Noland.[36] This suggests that this type of more detailed model can control for other elements of VMT growth. The work of Duranton and Turner, however, gives much higher elasticity estimates.[37] They use an approach that allows one to assess whether one can claim that building new capacity actually leads to more travel (i.e., that "A causes B").[38] While this model only uses cross-sectional data, it controls for endogeneity, a shortcoming of some of the other models.

Fulton et al. also controlled for endogeneity using county-level data.[39] A growth model is used to correlate VMT growth with increases in lane-miles, allowing them to conduct a test of whether capacity growth (i.e., lane-mile growth) precedes associated increases in VMT.[40] A growth variable for lane-miles is used as an instrument in a second stage to address any bias caused by VMT affecting growth in lane-miles (i.e., simultaneity bias) with one-year and two-year lag periods. As a result Fulton et al. report short-term lane-mile elasticities between 0.2 and 0.6, the first successful model to establish a causal linkage.

Cervero and Hansen demonstrate the mutual causality of VMT and lane-mile expansion.[41] Using simultaneous equations for both supply and demand and also instrumental-variable regressions, which include political variables as instrument, they find a statistically significant induced travel effect. They also find that increases in VMT lead to more road capacity, i.e., that there is a two-way effect confirming the view that planners have some foresight about where road capacity will be demanded. This latter effect, however, is smaller than the estimated coefficient on lane-miles, associated with VMT.

Facility-specific studies have examined the impact of specific links or corridors within the transportation network and subsequent growth in VMT. Cervero estimates a structured-equation model that includes changes in road speed, as the mediating influence on behavioral change, and links new capacity to development activity and VMT.[42] His estimates also account for endogenous effects of how VMT and development activity affect both speeds and increased road capacity. Road capacity is measured using a selection of specific projects, rather than aggregate changes in lane-miles as other models have done. He finds statistically significant effects, which suggests a useful approach for linking individual projects to potential increases in carbon emissions.

Cervero's results are listed in table 4.1 and provide both short- and long-run elasticity estimates.[43] While he claims that he shows smaller effects than other studies, this is not an accurate interpretation of other results. He attributes about 40 percent of VMT growth to capacity improvements, while Noland found at most 28 percent, with other demographic and economic factors associated with the remainder.[44] Despite this flaw in how Cervero's model results have been interpreted, the structure is useful for dissecting different sources of growth in VMT (speeds and development effects) and for accounting for endogeneity.

Models using disaggregate data also find similar results. Strathman et al. select a sample of roughly 12,000 individual respondents from the 1995 Nationwide Personal Transportation Survey (NPTS) and 48 urban areas from the Texas Transportation Institute database for their model estimation.[45] They report a direct elasticity of per-capita roadway capacity to VMT of 0.29; this means that for a 1 percent increase in lane-miles per capita there is a corresponding 0.29 percent increase in VMT. They also estimate an indirect elasticity (representing the secondary effects due to changes in land use) of 0.033 (correspondingly implying a 0.033 percent increase in VMT associated with a 1 percent increase in lane-miles per capita). The meaning of the indirect elasticity value is questionable because residential- and employment-location choices are affected by induced travel effects. Barr also used the NPTS data to estimate short-term elasticities.[46] However, his sample was nationwide and included roughly 27,000 households of which 61 percent were in urbanized areas and 63 percent had access to public transportation. A model was estimated with VMT as the dependent variable, while the independent variables included the inverse of speed, census tract population density, annual and per-capita family income, household size, number of workers in households, the median household income of the census tract, and an error term.

Barr estimates travel time elasticities between −0.3 and −0.5 (that is, a 1 percent increase in travel time reduces VMT by between 0.3 percent and 0.5 percent).

Another approach to evaluating induced travel effects is to use an integrated land-use/transport-demand model. Rodier et al. implement the MEPLAN integrated land-use/transport-model for metropolitan Sacramento to explicitly examine induced travel effects.[47] MEPLAN, in theory, allows for the full integration of land-use and economic effects associated with lane-mile increases, as opposed to the typical assumption that any land-use change is not affected by changes in the road network made by most travel demand models. They find significant differences associated with forecasts that assume induced travel effects compared to not using sufficient feedback. Rodier et al. estimate forecasts for 25- and 50-year predictions.

The MEPLAN modeling framework that Rodier et al. use includes separate but interactive land and transportation markets.[48] The region is disaggregated spatially and classified by land-use type. Discrete choice models predict the location choices based on the attractiveness of each, which is a function of activity-specific input costs including transportation costs based on a transportation network and location-specific disutilities. Through an incremental model, lags provide feedback of transportation costs from one period to the land market model of the next period, so that land use is handled dynamically. This application includes eleven industry classifications to match employment with locations; three classifications of household income that incorporate residential location; business consumption of household labor; business activities of households to purchase goods and services; and consumption of space based on elasticities for seven types of land use. Vacant land and different rents paid for similar land use are also tracked.

In addition to the land-use components, they also made substantial improvements

Table 4.2 Estimates using travel-demand models

Model	Method	Scale	Type	Long-term Elasticities
DeCorla-Souza (2000)				
No Feedback	Four step	Facility	Travel time	−0.7
Feedback	Four step	Facility	Travel time	−1.1
Rodier et al. (2001)				
25 years	MEPLAN	Metro	Lane-miles	0.8
50 years	MEPLAN	Metro	Lane-miles	1.1

to the Sacramento regional transportation model, thus better incorporating other induced travel effects (beyond the long-term land-use impacts). These specifically include better feedback from trip generation, distribution, mode choice, and network-assignment steps of the model.

Rodier et al. derive estimates that imply an increase in lane-miles of 1 percent will result in an increase in vehicle travel of 0.8 percent over 25 years and a larger increase of 1.1 percent over 50 years.[49] Table 4.2 summarizes this result and displays these as lane-mile elasticities. Using a simpler model without the intricate feedbacks of Rodier et al., DeCorla-Souza reports a similar result based on changes in travel time of a 0.7–1.1 percent increase in VMT for a 1 percent decrease in travel time (the latter with some feedback accounted for in the modeling, see table 4.2).[50] DeCorla-Souza does not state the time frame in connection with these travel-time elasticities.

Other approaches include those of Waddell et al. that implements the UrbanSim location-choice model.[51] These types of land-use models are dependent on large databases disaggregated to parcel-level data and are estimated using discrete-choice methods, which can introduce significant levels of uncertainty and error in the results. As an example of the inherent uncertainty of all these approaches, Rodier and Johnston found emissions forecasting to be very sensitive to population and employment growth such that it would likely swamp any measurable impact from a specific project, using regional modeling approaches.[52]

Implications for Sustainable Transportation Policy

This chapter has focused on establishing the basic theoretical features of how traffic is induced in response to new road capacity. Basic economic theory provides a fundamental relationship between road supply and demand, with travel time of individuals being the price that is determined at equilibrium. We further show how basic urban economic theory implies that long-run effects are captured by changes in land use, and consequently new development can occur in response to increased accessibility. Thus, this leads us to conclude that theoretically one cannot reduce congestion through new road projects, and building new roads that access undeveloped land will result in increased vehicle travel. Our review of empirical studies finds conclusive evidence of this theoretical relationship.

It is widely acknowledged that improvements to travel-demand modeling are needed to adequately assess both land-use impacts and the amount of travel by non-motorized modes. However, these methods still focus on evaluating the benefits of travel, rather than those associated with alternative development patterns. As noted above, benefits accrue to both current landholders and consumers; nothing need be said about mobility benefits if an urban economic framework is used. The key distinction in evaluation is that this type of access-based approach to assessment avoids the political argument that a given project will reduce congestion. Induced travel suggests that capacity expansions won't succeed at reducing congestion, but rather will succeed at spurring development. The political argument can then focus on how that development can occur. From a sustainable-transportation perspective, this helps to shift the debate to whether development should be dispersed or, on the other hand, made more concentrated and amenable to non-motorized modes of travel.

As cheap oil resources diminish, the consequences of dispersed development choices will become clear. More-expensive gasoline costs will lead consumers to prefer more-accessible locations compared to those that are more dispersed. It is necessary, of course, to make it easier for these accessible locations to provide more options for traveling without a car. Of course, more fuel-efficient vehicles and those using alternative sources of energy will also increase and may enable some dispersed location choices. However, many of these technologies are insufficient to meet long-term goals of greenhouse-gas reductions for avoiding drastic climate change, so more sustainable transportation will still require the development of more accessible locations with more choice of other modes of travel.

The question for policy makers is how to put this into practice. California provides a good example with the implementation of Senate Bill 375, which requires coordination of transportation and land-use planning and provides incentives for regions to target development toward "transit-rich" locations. While it is too early to gauge the success of this policy in practice, it does provide a useful way of thinking about better development choices. Investments should focus on increasing public transit options but also on providing the ability for easier (and safer) walking for local activities. New development should be concentrated and mixed with a multitude of economic activities, with the goal of providing a mix of residential choices for consumers.

Notes

1. Phil B. Goodwin, "Empirical Evidence on Induced Traffic," *Transportation* 23, no. 1 (1996): 35–54.

2. This is as opposed to statistical models that merely show that there is an *association* between two variables but are unable to claim that "A" *causes* "B."

3. Douglas B. Lee Jr., Lisa A. Klein, and Gregorio Camus, "Induced Traffic and Induced Demand," *Transportation Research Record: Journal of the Transportation Research Board* 1659, no. 1 (1999): 68–75; see also: Robert B. Noland and Lewison L. Lem, "A Review of the Evidence for Induced Travel and Changes in Transportation and Environmental Policy in the US and the

UK," *Transportation Research Part D* 7, no. 1 (2002): 1–26; Huw C. W. L. Williams and Yaeko Yamashita, "Travel Demand Forecasts and the Evaluation of Highway Schemes Under Congested Conditions," *Journal of Transport Economics and Policy* 26, no. 3 (1992): 261–82.

4. Patrick DeCorla-Souza, "Induced Highway Travel: Transportation Policy Implications for Congested Metropolitan Areas," *Transportation Quarterly* 54, no. 2 (2000): 13–30.

5. Todd Litman, "Generated Traffic: Implications for Transport Planning," *ITE Journal* 71, no. 4 (2001): 38–46; see also Noland and Lem, "A Review of the Evidence."

6. Goodwin, "Empirical Evidence on Induced Traffic"; see also: Phil B. Goodwin, "A Review of New Demand Elasticities with Special Reference to Short- and Long-Run Effects of Price Changes," *Journal of Transport Economics and Policy* 26, no. 2 (1992): 155–69; and Phil Goodwin, Joyce Dargay, and Mark Hanly, "Elasticities of Road Traffic and Fuel Consumption with Respect to Price and Income: A Review," *Transport Reviews* 24, no. 3 (2004): 275–92.

7. Lee, Klein, and Camus, "Induced Traffic and Induced Demand"; see also: Noland and Lem, "A Review of the Evidence"; Lewis M. Fulton, Robert B. Noland, Daniel J. Meszler, and John V. Thomas, "A Statistical Analysis of Induced Travel Effects in the US Mid-Atlantic Region," *Journal of Transportation and Statistics* 3, no. 1 (2000): 1–14; and Roger Gorham, *Demystifying Induced Travel Demand*, Sustainable Urban Transport Technical Document #1 (Eschborn, Germany: Deutche Gesellschaft, 2009).

8. Robert Cervero, "Transit Pricing Research," *Transportation* 17, no. 2 (1990): 117–39; see also: Michael A. Kemp, "Some Evidence of Transit Demand Elasticities," *Transportation* 2, no.1 (1973): 25–52.

9. Noland and Lem, "A Review of the Evidence."

10. *Exogenous* refers to an effect that is not internal to the system being studied; for example, population growth that is independent of any changes in the transportation system would be considered an exogenous impact on the system.

11. Noland and Lem, "A Review of the Evidence."

12. Yacov Zahavi and Antti Talvitie, "Regularities in Travel Time and Money Expenditures," *Transportation Research Record* 750 (1980): 13–19.

13. Yacov Zahavi and James M. Ryan, "Stability of Travel Components Over Time," *Transportation Research Record* 750 (1980): 19–26.

14. Noland and Lem, "A Review of the Evidence."

15. Anthony Downs, *Stuck in Traffic: Coping with Peak-Hour Expressway Congestion* (Washington, DC: Brookings Institution, 1992).

16. Noland and Lem, "A Review of the Evidence"; see also: Robert Cervero, "Induced Travel Demand: Research Design, Empirical Evidence, and Normative Policies," *Journal of Planning Literature* 17, no. 1 (2002): 3–20; Robert Cervero, "Road Expansion, Urban Growth, and Induced Travel: A Path Analysis," *Journal of the American Planning Association* 69, no. 2 (2003): 145–64; and R. Gorham, *Demystifying Induced Travel Demand*.

17. Robert B. Noland, "Transport Planning and Environmental Assessment: Implications of Induced Travel Effects," *International Journal of Sustainable Transportation* 1, no. 1 (2007): 1–28.

18. Robert B. Noland and William A. Cowart, "Analysis of Metropolitan Highway Capacity and the Growth in Vehicle Miles of Travel," *Transportation* 27, no. 4 (2000): 363–90.

19. Noland, "Transport Planning and Environmental Assessment."

20. Ibid.

21. Justin S. Chang, "Models of the Relationship between Transport and Land Use: A Review," *Transport Reviews* 26, no. 3 (2006): 325–50.

22. Litman, "Generated Traffic."

23. Noland, "Transport Planning and Environmental Assessment."

24. Daniel J. Graham, "Variable Returns to Agglomeration and the Effect of Road Traffic Congestion," *Journal of Urban Economics* 62, no. 1 (2007): 103–20.

25. Richard Arnott and Kenneth Small, "The Economics of Traffic Congestion," *American Scientist* 82 (1994): 446–46.

26. Cervero, "Transit Pricing Research"; Kemp, "Some Evidence of Transit Demand Elasticities."

27. Cervero, "Induced Travel Demand"; see also: Patrick DeCorla-Souza and Harry Cohen, "Accounting for Induced Travel in Evaluation of Metropolitan Highway Expansion" (report, US Department of Transportation, Federal Highway Administration, Washington, DC, 1998).

28. Cervero, "Induced Travel Demand"; see also: Lawrence C. Barr, "Testing for the Significance of Induced Highway Travel Demand in Metropolitan Areas," *Transportation Research Record: Journal of the Transportation Research Board* 1706, no. 1 (2000): 1–8.

29. Caroline J. Rodier, John E. Abraham, Robert A. Johnston, and John D. Hunt, "Anatomy of Induced Travel, Using an Integrated Land-Use and Transportation Model in the Sacramento Region" (preprint for the Annual Meeting of the Transportation Research Board, Washington, DC, 2001).

30. Mark Hansen, David Gillen, A. Dobbins, Yuanlin Huang, and Mohnish Puvathingal, "The Air Quality Impacts of Urban Highway Capacity Expansion: Traffic Generation and Land-Use Change" (report, Institute of Transportation Studies, University of California, Berkeley, 1993).

31. Mark Hansen and Yuanlin Huang, "Road Supply and Traffic in California Urban Areas," *Transportation Research Part A* 31, no. 3 (1997): 205–18.

32. This is known as "heteroskedasticity."

33. Kent M. Hymel, Kenneth A. Small, and Kurt Van Dender, "Induced Demand and Rebound Effects in Road Transport," *Transportation Research Part B: Methodological* 44, no. 10 (2010): 1220–41.

34. Robert Noland, "Relationships between Highway Capacity and Induced Vehicle Travel," *Transportation Research Part A* 35, no. 1 (2001): 47–72.

35. Their estimated model used a 3-stage least-squares regression, also known as a structural model.

36. Noland, "Relationships between Highway Capacity and Induced Vehicle Travel."

37. Gilles Duranton and Matthew A. Turner, "The Fundamental Law of Road Congestion:

Evidence from US Cities" (publication of the National Bureau of Economic Research, NBER Working Paper No. 2652, 2011).

38. They estimate an instrumental-variable model using historical measures of exploration routes, railroads, and the planned 1947 Interstate Highway System. The theory behind this approach is that it controls for factors that may influence the growth in capacity independent of current levels of vehicle travel.

39. Fulton et al., "A Statistical Analysis of Induced Travel Effects."

40. The procedure used is a Granger Causality Test.

41. Robert Cervero and Mark Hansen, "Induced Travel Demand and Induced Road Investment: A Simultaneous Equation Analysis," *Journal of Transport Economics and Policy* 36, no. 3 (2002): 469–90.

42. Cervero, "Road Expansion."

43. Ibid.

44. Noland, "Relationships between Highway Capacity and Induced Vehicle Travel."

45. James G. Strathman, Kenneth J. Dueker, Thomas Sánchez, Jihong Zhang, and Anne-Elizabeth Riis, "Analysis of Induced Travel in the 1995 NPTS" (report, Center for Urban Studies, Portland State University, 2000).

46. Barr, "Testing for the Significance of Induced Highway Travel Demand."

47. Rodier et al., "Anatomy of Induced Travel."

48. Ibid.

49. Ibid.

50. DeCorla-Souza and Cohen, "Induced Highway Travel."

51. Paul Waddell, Gudmundur F. Ulfarsson, Joel P. Franklin, and John Lobb, "Incorporating Land Use in Metropolitan Transportation Planning," *Transportation Research Part A: Policy and Practice* 41, no. 5 (2007): 382–410.

52. Caroline J. Rodier and Robert A. Johnston, "Uncertain Socioeconomic Projections Used in Travel Demand and Emissions Models: Could Plausible Errors Result in Air Quality Nonconformity?" *Transportation Research Part A* 36, no. 7 (2002): 613–31.

Bending the Curve

How Reshaping US Transportation Can Influence Carbon Demand

5

DERON LOVAAS AND JOANNE R. POTTER

Pulling back the lens from the memorable image of the broken pipe miles underwater spewing crude oil into a rich Gulf ecosystem brings into focus an energy crossroads ahead. Call it "extreme drilling"—the energy industry's analog to sports—which trades potentially high returns for skyhigh economic and environmental risks. This once-minor side of the energy industry is rapidly becoming mainstream, and it's worth asking why.

The short answer to the expansion of extreme drilling is the world's ever-climbing thirst for energy—and for transportation energy, in particular. Deep-water drilling can be viewed as part of what one analyst has termed the "deconventionalization" of oil production capacity and reserves necessary to meet current and projected demand for transportation fuel.[1] In just one day we will consume more than 84 million barrels of oil, which is about 3.5 billion gallons of the black stuff.[2] The lion's share goes into the fuel tanks of vehicles, boats, and aircraft, with the whole transportation sector accounting for about more than half of global demand.[3]

How can we "bend the curve" of the seemingly inexorable transportation demand for oil? While advances in fuel efficiency and new technologies offer hope, continuing growth in demand for carbon-intensive travel—compounded by rapid urbanization and population growth worldwide—present a daunting challenge. Yet recent studies indicate that a combination of strategies to shift travel activity could achieve more fuel-efficient mobility, complementing the essential progress we need to make in technology and the decarbonization of fuels. In this chapter, we lay out the current global projections for transportation energy consumption and the challenge confronting today's transportation leaders. We then discuss steps the US transportation community can take to reshape transportation in the coming decades, developing new examples of vibrant, mobile societies that can reduce our dependence on oil and promote a sustainable future across the globe.

Transportation's Demand for Oil Compounds the Global Energy Challenge

The Energy Information Administration at the Department of Energy recently unveiled its latest projections of energy consumption, showing that oil demand is projected to continue increasing about 15 percent in the next two decades (fig. 5.1). This rise—though significant—is lower than previous years' projections, due mainly to projected improvements in fuel efficiency as new performance standards raise the bar for grams of greenhouse-gas pollution emitted per mile (gpm) and miles per gallon of fuel used (mpg). Biofuels—mostly ethanol—are also projected to account for some of the liquid consumption, pulling the projected trend line closer to flat.

Overall US oil consumption is nearly 19 million barrels a day, taking in more than one-fifth of current global production. The International Energy Agency projects that, assuming implementation of various policy commitments to reduce heat-trapping emissions by saving oil, global consumption will rise to 99 million barrels daily by 2035—an eye-popping thousand barrels per second (86 million barrels daily),[4] well in excess of what one energy expert calls a "break point" for consumption. Nearly all the growth is from nonindustrialized nations, with about half coming from China alone.

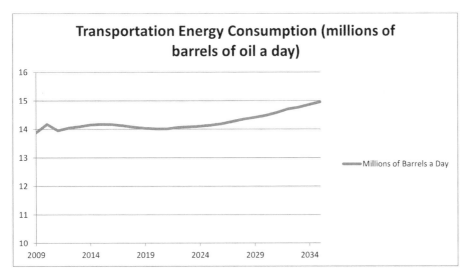

Figure 5.1   US transportation energy consumption. (Source: US Energy Information Administration.)

As billions of people in growing economies join the motorization trend, the world is moving inexorably to what two other authors have noted is the crucial "two billion car" mark.[5] In other words, even as the United States, Europe, and Japan trend lines flatten out (or decline—the IEA projects Europe's demand for oil will actually drop by 6 mbd by 2035), the developing world will more than make up for it.

This is already driving energy production in new directions, as companies and countries scour the Earth for new liquids to meet rising demand. One energy expert tells NRDC Executive Director Peter Lehner in his recent book about the Gulf oil spill, *In Deep Water*:

> "Why are they out there? They're there because that's where the resource is," said Robert Bryce of the Manhattan Institute. "For any of these huge companies to move the needle in any meaningful way, in terms of the resource base, they have to be looking for elephants."
>
> And the elephants are in deep water. . . .[6]

Unconventional oil—specifically from Canada's tar sands and Venezuela's extra-heavy oil, as well as liquefied coal and natural gas—is also projected to play a larger role by the 2030s.[7] Oil industry veteran Leonardo Maugeri says that there is a "process of 'deconventionalization' of reserves . . . taking place that will probably make the future supply of oil the result of a mosaic of many increments . . . from unconventional sources. . . ."[8] Currently, unconventional oil accounts for about 3 percent of global production, which IEA projects to expand to nearly 11 percent by 2035.[9] Despite this diversification of sources, however, the cartel OPEC still looms large, with IEA projecting its output to account for nearly one-half of global production by 2035.[10]

Overall, this is a dystopian energy future in the oil marketplace. As the experience in the Gulf of Mexico in 2010 shows, the energy game is becoming riskier and riskier. And, even more alarming, the carbon-intensity of securing this energy is increasing, both because it takes more energy to secure harder to reach energy sources, and because the extraction and refining of heavier, poor-quality oils, tar sands, oil shale, and liquefied coal is very carbon-intensive relative to conventional petroleum. All of this activity means a steady, alarming "carbon drift" in the energy industry. The bottom line is that the climatic and environmental costs and consequences are likely to be quite high.[11] And as energy expert Anne Korin has testified, OPEC's monopolistic practices in the oil marketplace undermine national security.[12]

Choosing a Preferable Path

There is an exit ramp from this future. Thanks to our monopsonistic status in the global marketplace, US action could have worldwide benefits as we develop—and export—technologies and techniques that save oil. We have a real chance to be an energy-secure beacon to the world.[13]

We have already seen that demand projections aren't destiny—the US government is proving this with historic increases in fuel-efficiency standards for cars and trucks, which are bending the projected US oil-demand curve for the first time in decades. Our success in technology advances has been, and will continue to be, vital to addressing the transportation-energy challenge. In fact, NRDC analysis shows that efficiency standards policy and increased use of biofuels—thanks in part to a renewable fuels standard—would cut oil demand by nearly 20 percent in just two decades.[14] The most visible evidence of this trend is, of course, the proliferation of hybrid-electric vehicles, with 38 (as well as seven pluggable electric vehicles) on the market currently and more to come in future model years.[15] In addition, pluggable cars are here, too, with 50 models expected to be available in five years. These include battery electrics such the Nissan Leaf and plug-in hybrid electrics such as the Chevrolet Volt.[16]

This new direction for fuel efficiency and an increasing number of fuel choices provide reasons for optimism. Is there more that can be done to tackle oil dependence, however? The answer is yes. More can be done, and more will be necessary. A complementary set of policy tools should be deployed nationally: travel efficiency and smart-growth strategies.

One of the first analytic steps toward fruitful use of any policies and techniques for influencing energy use is an assessment of what is technically achievable, assuming their deployment.[17] This was a key reason we convened a broad group of stakeholders to steer a new study, commissioned by the respected consulting firm Cambridge Systematics.[18] The final product, unveiled in 2009, is entitled *Moving Cooler: An Analysis of Transportation Strategies for Reducing Greenhouse Gas Emissions*. While the study, as the title says, focuses on reducing heat-trapping pollution, the means for doing so is by calculating fuel savings, so it also serves as an assessment of the potential to reduce oil dependence.

The study examined the effectiveness of about 50 measures that fit into nine categories:

1. Pricing (e.g., tolls, pay-as-you-drive insurance, VMT fees, carbon/fuel taxes);
2. Land use and smart growth;
3. Nonmotorized transportation (i.e., walking and biking);

4. Public transportation improvements;
5. Regional ride-sharing and commute measures;
6. Regulatory measures (e.g., reduced speed limits, which actually proved quite effective);
7. Operational/ITS strategies;
8. Capacity/bottleneck relief; and
9. Freight sector strategies.

The aggregated results of the analysis are impressive. To understand them it is important to know that the study built three scenarios that vary based on aggressiveness of implementation (i.e., variable assumptions of how extensive geographically, how rapidly, and how intensively each scenario would be implemented). And then the study mapped out emission-reduction potential for six different thematic "bundles" that could be deployed depending on the capacity and priorities of implementing institutions:

1. Near-Term/Early Results;
2. Long-Term/Maximum Results;
3. Land Use/Transit/Nonmotorized Transportation;
4. System and Driver Efficiency;
5. Facility Pricing; and
6. Low Cost.

Figure 5.2 shows technically achievable fuel savings under the maximum deployment (i.e., most aggressive) implementation scenario for each of the bundles, excluding the possible overlay of an economy-wide price change using a policy tool such as a gas tax.

One of the conclusions that can be drawn from the *Moving Cooler* analysis may seem obvious but it's worth stating: While no one strategy will "solve" the transportation fuel challenge, linking multiple strategies can add up, and can afford synergies as well. Thoughtful combinations of strategies can generate not just fuel savings, but multiple environmental, social, and economic benefits. Another paper found that providing transportation alternatives also helps make pricing more politically viable and therefore durable as a policy.[19]

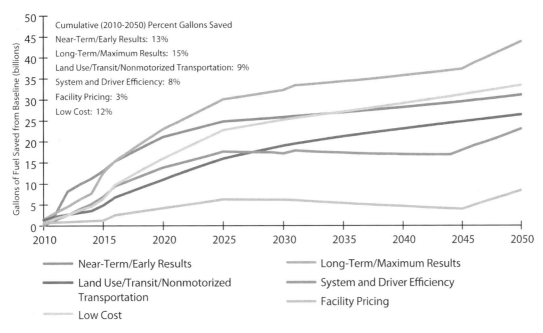

Figure 5.2   Gallons of fuel saved by each bundle at maximum level of deployment (without economy-wide pricing). Note: The figure displays the number of gallons of fuel saved from the baseline for each bundle, assuming maximum level of deployment during the 2010–2050 time period. The cumulative percentage of gallons saved during this time period ranges from a high of 15 percent saved from the Long-Term/Max Results bundle to a low of 3 percent saved from the Facility Pricing bundle. (Source: reprinted courtesy of the Urban Land Institute and *Moving Cooler: An Analysis of Transportation Strategies for Reducing Greenhouse Gas Emissions*.)

The Importance of Pricing

The *Moving Cooler* study confirmed what others have found—namely, that pricing measures can be the most effective means to save fuel. We examined ten means to alter the price of transportation goods, in the categories of parking fees and permits, road pricing, pay-as-you-drive insurance, and economy-wide mileage and carbon fees. The results were quite dramatic, as shown in figure 5.3. For all the measures other than pricing, the maximum potential is 24 percent reduction in emissions compared to the baseline, with pricing added it is 35 percent. To be clear, this would entail ramping up to European-level fuel surcharges, which is exceedingly unlikely in the United States.

While pricing approaches are in theory highly efficient in dampening fuel consumption, in reality the political barriers are daunting: Lawmakers view hiking gas

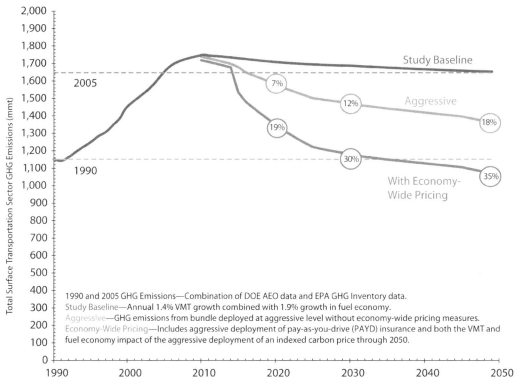

Figure 5.3   Total surface transportation sector GHG emissions (mmt). (Source: reprinted courtesy of the Urban Land Institute and *Moving Cooler: An Analysis of Transportation Strategies for Reducing Greenhouse Gas Emissions*.)

prices as the untouchable "third rail" of public policy. Beyond this aversion, however, lies a serious concern about the equity effects of pricing: the impact of higher fuel costs on lower-income populations. In 2008, Americans with the lowest household incomes—the lowest 20 percent of families—paid 9.9 percent of their monthly income on gasoline. This compares to only 3.9 percent for wealthier Americans.[20] In order for pricing approaches to be fair, these policies need to be packaged with policies ensuring that poor citizens can have equitable access to jobs and community life.

A variety of pricing tools are being used to varying degrees across the country. For example, high-occupancy toll (HOT) lanes have been deployed thus far to good effect in California, Colorado, Florida, Minnesota, and Washington.[21] In 2003, two transportation experts proposed moving up to a system approach when applying this technology, so-called HOT Networks, in eight large metropolitan areas.[22] The proposal would combine the technique with another innovation—bus rapid transit (BRT). In fact, providing ample transportation alternatives such as BRT is important in address-ing inequitable effects of pricing. As we wrote in *Moving Cooler*:

> Because lower-income people rely more on public transportation than other groups, public transportation improvements can potentially channel higher per-centages of benefits to lower-income people and those without other mode choices, such as people who reside in rural areas.[23]

Another pricing strategy examined by *Moving Cooler* is auto insurance reform, whereby this cost is made variable based on mileage rather than a fixed annual fee. Under the current fixed-rate system, high-mileage drivers are essentially cross-subsi-dized by low-mileage ones. Attaching the price of insurance to mileage driven would remedy this inequity, saving most consumers money, and saving a great deal of fuel as well by providing drivers an incentive to economize on travel. A Brookings study found that two-thirds of all households would save an average of $270 per auto annu-ally by shifting to mileage-based insurance fees.[24] And a recently released study from the Mobility Choice coalition found potential fuel savings of 56 million barrels yearly by 2020.[25]

As figure 5.3 shows, by far the biggest potential fuel-saver is an economy-wide fee of some sort, whether on carbon, fuel, or mileage. A fee on fuel or carbon would have an especially large effect, since it would both depress travel demand and stimulate advances in fuel efficiency. In fact, *Moving Cooler* found that annual greenhouse-gas emissions could be reduced by 37–325 million metric tons by using such an economy-wide pricing policy.[26] And a more recent assessment concluded that a 25¢ per-gallon fee would save nearly 1.3 million barrels of oil daily by 2030, or about 7 percent of cur-rent consumption.[27]

One of the biggest challenges with instituting such a fee, of course, is political vi-ability. Congress hasn't seen fit to increase the federal gasoline tax in almost 20 years, despite crumbling infrastructure and at least $35 billion in general fund transfers to the Highway Trust Fund to maintain its solvency.[28] However, there is a solid case for adopting such a fee based on the salutary effects, from increased energy independence to a reduction in our massive trade imbalance—oil has accounted for as much as half the trade deficit in the 2000s—to desperately needed infrastructure investments. It would also align us with the rest of the industrialized world, as figure 5.4 shows.

The Need for Smarter Land Development

Adequate pricing is necessary but insufficient for easing the pressure to scour the Earth for more hydrocarbons, however. We must also address land-development

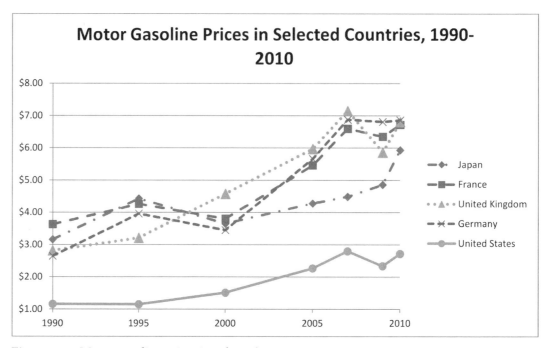

Figure 5.4   Motor gasoline prices in selected countries, 1990–2010.

issues, and specifically the suburban sprawl that dominates the marketplace despite shifting demand. This is a long-term strategy due to long implementation time frames (requiring, for example, changes to planning and zoning) with a large potential payoff by 2050.[29]

Land-use changes that tend to reduce vehicle-miles of travel, thus saving fuel, are often described as "smart growth." There are various definitions of smart growth, including consensus principles of the Smart Growth Network, which includes a range of groups from the Association of Metropolitan Planning Organizations to the National Association of Realtors to the US EPA. According to this group (which also includes NRDC), in principle, smart growth should:

- Utilize compact building design;
- Create a range of housing opportunities and choices;
- Create walkable neighborhoods;
- Encourage community and stakeholder collaboration;
- Foster distinctive, attractive communities with a strong sense of place;
- Make development decisions predictable, fair, and cost-effective;
- Mix land uses;
- Preserve open space, farmland, natural beauty, and critical environmental areas;

• Provide a variety of transportation choices; and

• Strengthen and direct development toward existing communities.[30]

This is a rather broad list, encompassing much more than land use. Existing literature on the subject has encouraged specific influences that happily all begin with the letter "D": Density, Diversity (i.e., a mix of uses such as commercial and residential), Design, Destination accessibility (i.e., how easily accessible trip attractions are), and Distance to transit.[31] Two scholars examined a sample of the most recent literature on these influences and found that while "relationships between travel variables and built environmental variables are inelastic . . . the combined effect of several built environmental variables on travel could be quite large."[32] In other words, dialing up changes to several or all of the Ds in metropolitan areas could have a large effect on nationwide vehicle-miles of travel. By *Moving Cooler*'s conservative analysis, as mentioned above, "large" means 4.4 percent greenhouse-gas emission reduction below the national baseline by 2050.

Another respected study (using somewhat different assumptions) determined that there is even greater potential. *Growing Cooler* found a potential 30 percent drop in vehicle-miles of travel assuming much more compact development by 2050, yielding a 7–10 percent decrease in emissions by 2050 (about twice that described in *Moving Cooler*).[33] This is not difficult to imagine if, as one of the authors has noted, a host of factors (e.g., demographic changes and increasing competition for hydrocarbon energy) mean that development 40 years hence is likely to be as different as today's development is from 1970.[34]

In fact, as other experts point out, the differences could be even starker due to a "demographic convergence," namely the aging of the Baby Boomers and the rise of the Millenials. These groups comprise nearly half of the American population, or more than twice the percentage of Americans—returning veterans and their spouses—who drove the breathtaking suburban explosion after World War II.[35] Market research shows that both demographic cohorts prefer smart-growth-style development. As one well-known publication predicts, this has the potential to change the game:

Next generation projects will orient to infill, urbanizing suburbs, and transit-oriented development. Smaller housing units—close to mass transit, work, and 24-hour amenities—gain favor over large houses on big lots at the suburban edge. People will continue to seek greater convenience and want to reduce energy expenses. Shorter commutes and smaller heating bills make up for higher infill real estate costs.[36]

What do such developments look like? Fortunately, we have a growing number of examples, many of which have been described by NRDC colleague Kaid Benfield in recent publications. Take Highlands Garden Village, the redevelopment of a 27-acre site in Denver, Colorado (fig. 5.5).[37] This site originally hosted an amusement park built in 1890, complete with zoo animals, rides, and a theater. The theater has been restored and an old carousel has been converted to a popular pavilion. The development offers housing options for a variety of income levels among its 306 homes as well as 75,000 square feet of commercial space. It is very walkable, with a variety of shops and restaurants nearby, and is bordered on the north by a convenient bus line.

Suburbs can also be retrofitted with smart-growth projects, although this can be more challenging. For example, the town of Reston, Virginia, 18 miles outside of Washington, DC, has many of the characteristics of sprawl, including cul-de-sacs, separated residential and commercial land uses, and streets without bike lanes or sidewalks as well as a lack of adequate public transportation.[38] However, Reston's development trajectory is bookended by smart-growth ideals. It was started in the 1960s by a visionary founder (Robert E. Simon, for whom REStton is named), who dreamed

Figure 5.5   Highlands Garden Village mixed-use development, Denver, Colorado. (Source: courtesy of Perry Rose.)

up a "satellite city" with seven small European-style villages, open to all races and incomes, with mixed land uses and a distinct town center. However, he had to sell the development rights, and by the 1980s Reston was as much sprawl as smart.

Then something remarkable happened—planning began for a new town center, dubbed (not very creatively) Reston Town Center. This development, begun in 1988, has expanded dramatically and now includes an open-air skating rink, more than 50 retail shops, 20 restaurants, a movie theater, and ample office space as well as a slew of townhouses, condos, and apartments. The DC-area Metro is building a new Silver Line out to Dulles Airport, past Reston, so soon the Town Center will be within biking, bus, and perhaps even walking distance of a major rail line.

On a household level, the energy savings of such development types compared to conventional sprawl are large—even more so, if coupled with greater vehicle and building efficiency. A recent EPA study produced a graph (fig. 5.6) based on national averages—consequently masking what is likely to be substantial regional variation but still useful for illustrative purposes—which is reproduced here.

Figure 5.6   Location efficiency: comparing household and transportation energy use by location. (Source: courtesy of Jonathan Rose Companies, LLC, with support from US EPA, 2010.)

The contrast between the bars bracketing the graph—the left transportation bar alone is nearly six times taller—is especially striking. Clearly, coupling smart growth with improved technology can make a big dent in our oil use.

Delivering Real Mobility Choices

This potential can't be realized without an array of adequate transportation options. Ideally, a traveler would have ample choices for getting from point A to point B. The hierarchy would include, beginning with the simplest and moving to the most costly and elaborate:

- Nonmotorized options—i.e., walking and biking
- Efficient loading of autos via carpooling, carsharing, and possibly even taxis
- Vanpools, shuttles, and jitneys
- Buses and BRT
- Intracity rail (e.g., light rail or subways)
- Intercity rail (including high-speed options)

Moving Cooler found that, individually, these options—with the exception of an aggressive set of strategies aimed at discouraging solo-driver commuting—would have a rather modest effect on fuel use and therefore emissions. However, when combined and coupled with land-use changes (in a "Land Use/Transit/Nonmotorized Transportation" bundle) the emission reduction potential jumps to a maximum of 15 percent by 2050.[39]

Setting land use aside, the biggest-ticket item by far on the list of strategies analyzed in this bundle is urban transit expansion. Another recent study confirms that even the relatively modest transit systems already in existence save fuel on a daily basis—a not-insubstantial 339,000 barrels daily (saving as much as was spilled in the Gulf last year—about 5 million barrels—in about two weeks).[40]

The biggest determinant of fuel savings in public transportation is load factor. When a bus runs empty or near-empty, those passengers might as well be driving. Figure 5.7 shows the differences in load factor for different modes of transportation.[41] Thankfully, a recent survey of 41 transit agencies found that most of them have initiatives to boost load factor, including "transit marketing campaigns, providing real-time transit information and trip planning software, and making improvements to transit stations and shelters. . . ."[42] In fact, a bigger problem for transit agencies may be an inability to keep up operationally with increasing loads, given budget cuts at the local and state levels.

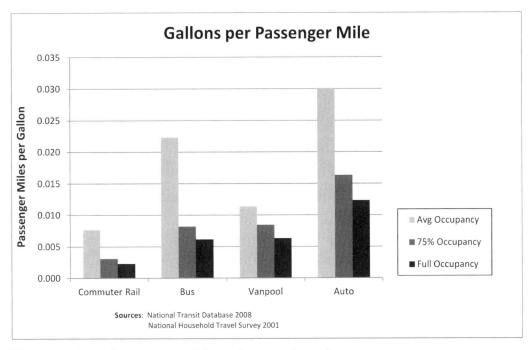

Figure 5.7   Load factor compared for various transit modes.

Public transportation has benefited from an enormous number of recent innovations. For example, Portland, Oregon, has long been a model city due to its remarkable land-use policies and commitment to transit. By expanding access and mobility options, it has even decoupled growth in vehicle miles of travel from economic growth, achieving a drop in miles of travel of 6 percent since 1990 while prospering, in contrast with the average for all US cities—a 10 percent increase in miles of travel.[43] In 2008, the city reestablished itself as a national leader in transit by manufacturing the first US-built modern streetcar. Establishing a domestic supply chain for new transit infrastructure and vehicles is key to creating domestic jobs and boosting the local economy.

Meanwhile, further south in Eugene, Oregon, the choice is BRT, not light rail. The EmX bus travels along two major corridors in this mid-sized city, and features many of the usual distinct, advantageous BRT characteristics:

- Exclusive right-of-way along most of the route
- Priority signalization at stop lights
- Fare collection at the stop, not aboard the bus

• Sleek station designs
• Frequent service

The system is also integrated with a network of bike paths. Ridership has soared, vastly exceeding projections, and more bus lanes and shelters are under way to build on this remarkable success.

While nonmotorized transportation's direct contribution to emission-reduction potential in *Moving Cooler* is small (no more than 0.5 percent below baseline), investments to expand these emission-free modes of travel "can be achieved at relatively low implementation cost, and with positive public health benefits."[44] And an assessment by two advocacy groups claims that the oil savings from current walking and biking—due to forgone auto trips—is more than 91,000 barrels daily; if the share of travel made through these modes increased from 9.6 percent to 25 percent that oil savings would jump to about 672,000 barrels a day.[45]

As with other strategies, synergies can be exploited by creating robust intermodal connections that make it easy for the public to walk and bike. Simple examples include bike lockers at commuter rail stations and bicycle racks on buses. Much more could be done, however. Boulder County in Colorado, for example, has a "Final Mile Initiative," a bicycle adoption program that provides commuters with both a bike and a locker at particularly busy bus stops.[46] Bicycles can be serviced at a local shop courtesy of the county.

Boulder is on the right track. What we need now is a proliferation of such innovative, robust investments in an array of more fuel-efficient transportation choices—rail, bus, biking, and walking—if we are to break our oil addiction and ease development pressure on places like the Gulf, the Arctic, and other fragile ecosystems.

A Brighter Future?

One of the biggest advantages of many of the strategies examined in *Moving Cooler* is that they offer flexibility to consumers. This is important to the goal of genuine energy independence or security. Increased efficiency in our vehicles and our buildings is absolutely necessary as well. These steps will help to drive down the oil intensity of the US economy (oil intensity is the ratio of oil used per dollar of gross domestic product). So will delivering more fuel choices (e.g., pluggable cars that can run on oil-free electricity *or* gasoline) and more mobility choices such as those described above.

However, fuel and mobility choices don't just decrease oil intensity over the long run, insulating our economy from price increases and lowering oil imports. They also allow us to "save oil in a hurry," to borrow the title of a thought-provoking International Energy Agency publication written by two respected analysts.[47] This is important in the case of more sudden price spikes, a real possibility in a volatile energy marketplace with a cartel en route to controlling more than half the world's supply of oil.

A sense of urgency about deploying techniques and strategies studied in recent technical analyses such as *Growing Cooler*, *Moving Cooler*, and the Transportation Research Board's *Driving and the Built Environment* is therefore warranted. The good news is that moving to a more secure and environmentally sound energy future can bring with it a revitalized economy, expanded consumer choice, and vibrant communities. The United States has the opportunity to lead the globe into a sustainable decarbonized future, with the transportation community playing an instrumental role. For the sake of our energy future, we hope we move expeditiously and we look forward to achieving this vision.

Notes

1. Leonardo Maugeri, *The Age of Oil: The Mythology, History, and Future of the World's Most Controversial Resource* (Westport, CT: Praeger, 2006).

2. Energy Information Administration, "Petroleum Statistics" (report, Energy Information Administration, Washington, DC, 2011), www.eia.gov/energyexplained/index.cfm?page=oil_home#tab2.

3. Energy Information Administration, "International Energy Outlook 2010: Transportation Sector Energy Consumption" (report, Energy Information Administration, 2010, www.eia.doe.gov/oiaf/ieo/transportation.html.

4. P. Tertzakian, *A Thousand Barrels a Second: The Coming Oil Break Point and the Challenges Facing an Energy-Dependent World* (New York: McGraw Hill, 2006).

5. D. Sperling and D. Gordon, *Two Billion Cars: Driving Toward Sustainability* (Oxford: Oxford University Press, 2009).

6. P. Lehner and B. Deans, *In Deep Water: The Anatomy of a Disaster, the Fate of the Gulf, and Ending Our Oil Addiction* (New York: The Experiment, 2010).

7. Energy Information Administration, "International Energy Outlook 2010: Liquid Fuels" (report, Energy Information Administration, 2010, www.eia.doe.gov/oiaf/ieo/liquid_fuels.html.

8. Maugeri, *The Age of Oil*, 220.

9. International Energy Agency, *World Energy Outlook 2011* (Paris, France: OECD/IEA, 2011).

10. Ibid.

11. D. Lovaas, "Taking the High Road to Energy Security," in *Business: The Magazine for Creating Sustainable Enterprises & Communities* 28, no. 3 (May–June 2006): 28.

12. A. Korin, "Rising Oil Prices, Declining National Security," Hearings of the US House of Representatives Committee on Foreign Affairs, May 22, 2008.

13. Monopsony is simply the flip side of monopoly; the former applies to buyers while the latter applies to sellers. In the oil marketplace, OPEC has monopolistic clout since it controls 40 percent of the supply and is the single largest seller; the United States arguably has monopsonistic influence since we account for more than one-fifth of global demand and are the single largest buyer.

14. L. Tonachel, "Moving Beyond Oil and Cutting Global Warming Pollution from Vehicles" (Natural Resources Defense Council report, November 30, 2010), www.e2.org/ext/doc /Tonachel%20Oil%20Forum%20Boston%202010–11–30.pdf.

15. See: HybridCars.com, www.hybridcars.com/news/june–2012-dashboard–47943.html, accessed July 19, 2012.

16. Ibid.

17. C. Cleveland and Christopher Morris, *Dictionary of Energy* (Oxford: Elsevier, 2006).

18. Moving Cooler sponsors: American Public Transportation Association, Environmental Defense Fund, Federal Highway Administration, Federal Transit Administration, Intelligent Transportation Society of America, the Kresge Foundation, Natural Resources Defense Council, Rockefeller Brothers Fund, The Rockefeller Foundation, Shell, Surdna Foundation, Urban Land Institute, and the US Environmental Protection Agency.

19. L. Munnich, "Enhancing Livability and Sustainability by Linking Congestion Pricing with Transit" (Transportation Research Board Paper 10–2492, November 14, 2009).

20. J. Potter et al., *Moving Cooler: An Analysis of Transportation Strategies for Reducing Greenhouse Gas Emissions* (Washington, DC: Urban Land Institute, July 2009).

21. Munnich, "Enhancing Livability."

22. R. Poole and K. Orski, "HOT Lanes: A Better Way to Attack Urban Highway Congestion," *Regulation* 23, no. 1 (2000): 15–20.

23. Potter et al., *Moving Cooler*, 74. Regarding discounted fares for low-income groups, I am part of a coalition called Mobility Choice, which advocates for transit vouchers to reduce inequitable effects due to higher fuel and road-use fees; these vouchers would also have other useful effects such as creating an increased opportunity to charge higher transit fares, thereby increasing farebox recovery and reducing transit subsidies, as well as a modest but measurable savings in fuel due to higher transit ridership.

24. J. Bordoff and N. Pascal, "Pay-As-You-Drive Auto Insurance," *Weekly Policy Commentary* (report, Resources for the Future, 2008), www.rff.org/Publications/WPC/Pages/12_15_08 _pay-as-you-drive_insurance.aspx.

25. A. Korin and D. Lovaas, *Taking the Wheel: Achieving a Competitive Transportation Sector Through Mobility Choice* (Washington, DC: Mobility Choice, 2010).

26. Potter et al., *Moving Cooler*, 44.

27. Korin and Lovaas, *Taking the Wheel*, 13.

28. The federal fuel excise tax was last increased in 1993. The $35 billion is relevant figure for 2008–2010, but it excludes spending in the American Recovery and Reinvestment Act (ARRA), which would effectively double the figure, and added to this total are billions of dollars transferred to finance the modest growth in the new transportation law (Making Progress for the 21st Century or MAP–21, P.L.112–41) . The federal transportation program has become a minor contributor to the nation's deficit and debt due to the lack of political will to increase the gas tax.

29. Potter et al., *Moving Cooler*, 42.

30. Smart Growth Network, "Smart Growth Principles" (*This Is Smart Growth* report, EPA and International City-County Management Administration, Washington, DC, 2006), www .smartgrowth.org/why.php.

31. R. Ewing and R. Cervero, "Travel and the Built Environment," *Journal of the American Planning Association* 76, no. 3 (May 11, 2010): 1–30.

32. Ibid.

33. R. Ewing et al., *Growing Cooler: The Evidence on Urban Development and Climate Change* (Washington, DC: Urban Land Institute, 2008).

34. R. Ewing, A. C. Nelson, and K. Bartholomew, *Response to Special Report 298* (Salt Lake City: University of Utah, 2009).

35. P. Doherty and C. Leinberger, "The Next Real Estate Boom: How Housing (Yes, Housing) Can Turn the Economy Around," *Washington Monthly* 42 (November/December 2010): 22–25.

36. Urban Land Institute and PricewaterhouseCoopers, *Emerging Trends in Real Estate 2010* (Washington, DC: Urban Land Institute, 2010).

37. The description of this remarkable project is drawn entirely from two of Kaid's blog entries: K. Benfield, "A close look at a smart growth icon: Denver's Highlands Garden Village," switchboard.nrdc.org/blogs/kbenfield/a_close_look_at_a_smart_growth.html, July 20, 2010.

38. F. K. Benfield, J. Terris, N. Vorsanger, *Solving Sprawl: Models of Smart Growth in Communities Across America* (New York: Natural Resources Defense Council, 2001).

39. Potter et al., *Moving Cooler*, 54.

40. L. Bailey, P. L. Mohktarian, PhD, and A. Little, *The Broader Connection between Public Transportation, Energy Conservation and Greenhouse Gas Reduction* (Fairfax, VA: ICF International, February 2008).

41. Korin, "Rising Oil Prices," 7.

42. F. Gallivan, *Current Practices in GHG Emissions Savings from Transit* (CD-ROM, Transportation Research Board, 2010 Annual Meeting).

43. E. Burgess and A. Rood, *Reinventing Transit: American Communities Finding Smarter, Cleaner, Faster Transportation Solutions* (New York: Environmental Defense Fund, 2009).

44. Potter et al., *Moving Cooler*, 42.

45. T. Gotschi, PhD, and K. Mills, JD, *Active Transportation for America: The Case for Increased*

Federal Investment in Bicycling and Walking (Washington, DC: Rails-to-Trails Conservancy, 2008).

46. K. J. Krizek and E. W. Stonebraker, *Bicycling and Transit—A Marriage Unrealized* (CD-ROM, Transportation Research Board, 2010 Annual Meeting).

47. R. B. Noland, W. A. Cowart, and L. Fulton, *Travel Demand Policies for Saving Oil During a Supply Emergency* (CD-ROM, Transportation Research Board, 2005 Annual Meeting).

Part 2

Transportation and Oil Dependence: A Modal Analysis

6 Public Transportation as a Solution to Oil Dependence

Bradley W. Lane

Gasoline Prices to Sustainability in Transportation

Hurricane Katrina's landfall on the Gulf Coast was a stark reminder of the American dependence upon petroleum for transportation. Among the many deep and significant impacts following the disaster was a dramatic spike in domestic gasoline prices. Data in figure 6.1 indicate that average retail gasoline prices in the United States rose by 20 percent in a week after the hurricane, and the timing of the disruption at the end of the summer peak travel season further emphasized the importance of the price of gasoline to most Americans.

Though the cost of gasoline constitutes one of the most notable observations in the American public consciousness, long-term price behavior for gasoline can be misunderstood. After the OPEC-induced oil shocks of the 1970s caused gasoline prices to spike, prices then stayed remarkably constant for nearly two decades. Starting in the late 1990s, prices began to exhibit considerable escalation and fluctuation. Data in figure 6.2 from the US Department of Energy show an increase of over 300 percent in the retail price of gasoline between 1999 and 2008. After a correction with the 2008 Great Recession, gasoline prices have since resumed an upward trend.[1]

The United States is responsible for over 20 percent of global oil demand, and transportation accounts for two-thirds of this demand.[2] Personal transportation in the United States is overwhelmingly dependent upon automobile travel; the vast majority of all passenger miles traveled in the United States are via the automobile,[3] and automobile vehicle-miles traveled (VMT) per capita in the United States is the highest in the world.[4] Automobile travel is a key contributor to the unsustainability of

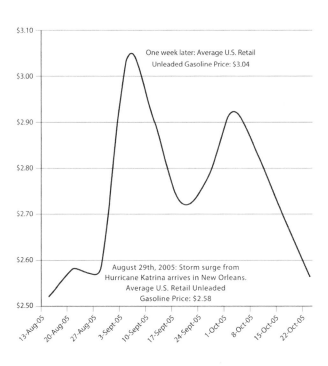

Figure 6.1   Gasoline price spikes following Hurricane Katrina. (Source: Energy Information Administration, "Weekly US Regular Conventional Retail Gasoline Prices" [US DOE, 2011], www.eia.doe.gov /petroleum/data_publica tions/wrgp/mogas_his tory.html, accessed June 25, 2012.)

Figure 6.2   Historical US gasoline prices, 1990–2011. (Source: Energy Information Administration, "Weekly US Regular Conventional Retail Gasoline Prices" [US DOE, 2011], www .eia.doe.gov/petroleum /data_publications/wrgp /mogas_history.html, accessed June 25, 2012.)

transport, largely due to its exclusive reliance on petroleum-based energy sources. Reducing VMT is thus key to alleviating US oil dependence.[5]

Increasing gasoline prices could help decrease automobile VMT. Though the car has been historically prioritized in transportation policy, development of roadway infrastructure, and land uses that facilitate automobile use, gasoline prices represent a potentially powerful market tool that can dictate a demand for change. Thirteen years of increasing fuel costs may indicate such a demand in the transportation marketplace. When we focus on reducing automobile VMT, however, we must recognize that there is a limit to how much travel can be reduced. Trips must still be made to work and to school, as well as for shopping and discretionary or leisure purposes. Changes to the built environment to reduce trip distances are slow and limited in deployment and effectiveness.[6] These factors make the demand for transportation highly inelastic.

This means we must consider displacing automobile travel with travel on other modes that are not as dependent upon oil for energy. This is where the role of the subject of this chapter, public transportation, comes into play. Walking, biking, and public transportation represent alternative modes to the automobile that, compared to other developed countries, are relatively unutilized here but could absorb displaced automobile travel.

Currently much of the transportation infrastructure and built environment in the United States is not conducive to these alternative travel modes. Many origins and destinations are too far apart to accommodate the slower speeds of walking and biking within the time budget of travelers. Even where this is not the case, there is often no infrastructure supporting these modes, making it impossible to walk or bike without violating traffic laws or subjecting oneself to physical danger.[7] Public transportation offers existing infrastructure support and the potential for higher speeds, which might better suit replacement of some automobile trips abandoned due to higher gasoline prices and can also chain trips together with other modes.

There are several benefits from transit trips displacing automobile trips. Transit vehicles can handle higher load factors than cars, and existing service features significant amounts of unused transit vehicle capacity in many places. Though currently the net amount of energy for an individual using transit is only slightly less than that of an individual using a car, more people riding each bus would decrease the net amount of energy required for this travel.[8] The mass-transit fleet is also further along in alternative fuels than the automotive fleet. At the time of this writing, the only alternative-fueled vehicles to be mass-marketed are the electric Nissan Versa and Chevy Volt. And though hybrids are growing in popularity, they still capture only a limited amount of the automotive market share and many do not provide much improvement in gasoline consumption over conventional gasoline vehicles.[9] Meanwhile, many buses run on compressed natural gas, electricity, or some kind of hybrid technology, and most of the intra-city rail service in the United States runs on electricity. Therefore, public transit may hold significant unrealized potential for alleviating unsustainability in transportation.[10]

Landscape of Public Transportation in the United States

Americans do not often consider public transportation as a viable mode of travel, and nationally the number of people who take transit to work is exceeded by the number driving alone, carpooling, walking, and biking.[11] While approximately 2 percent of all passenger trips occur on transit, just over 1 percent of passenger-miles traveled are on transit, indicating that transit trips are not only less frequently taken but also shorter than automobile trips.[12] The bulk of these trips occur in the urban agglomerations of the Northeast, large cities of the West Coast, and a couple of large metropolitan areas in the central United States.[13] Though the United States once boasted the most extensive streetcar system in the world as well as impressive intercity rail service, much of this was gone by the late 1950s.[14] Private-sector, for-profit transit operators disappeared and were replaced by subsidized public transportation. Many American cities operate transit as a social service for providing access to major activity nodes in the city for those who do not own automobiles.[15] This population is often referred to as "transit captive" and is made up mainly of low-income, central-city residents of urban areas.[16] Data from the American Public Transportation Association indicate that the aggregate number of passenger trips on transit in the United States declined by 50 percent by the early 1970s, and by 1995 was still below the level of ridership seen immediately post–World War II, despite national population growth of roughly 100 million people.[17]

Urban form and public policies of the last half of the twentieth century have also contributed to a difficult environment for transit. While the United States is more urban and less rural than it was in transit's heyday, the cities in which transit operates have declined in density and instead adapted to a more auto-oriented suburban form. Cheap gasoline with minimal taxes, preferences for single-occupancy detached-home ownership, and massive investment in roadway construction leaving few disincentives for driving, as well as great distances between origins and destinations, and poor access for walking and biking all conspire to make the United States "the developed world's most unfavorable public transport environment."[18]

There are many reasons behind this. The United States does not face nearly the same space pressures that most of the developed world does, and because of its relative newness as a country, there are fewer of the denser, older, historical sections of cities to be preserved and to direct urban form. The surge in wealth from increased production during and after World War II in the United States also contributed to the development of the idea of home ownership as an economic ideal and automobile ownership as a further achievement of the American dream.[19] However, a host of major federal policy initiatives, including and related to the Federal Housing Act of 1934 and the Interstate Highway Act of 1956, also (literally) paved the way for the decline of transit and the institution of driving as the dominant mode of travel in the United States. Policies at multiple levels of government continue to focus on automobile facilitation exclusive to other modes, and thinking about transport in the United States centers almost exclusively on the automobile.

Despite these long-standing trends and entrenched ideological and policy facilitators for automobile use, in the last few decades transit has experienced a slow but steady revival. Conventional bus service has been improved in many cities, and new modes of transit such as express bus and light rail have enhanced the serviceability and image of transit operations. Data from the American Public Transportation Association, shown in figure 6.3, indicates that mass transit has been increasing in aggregate counts of unlinked passenger trips in the United States since the mid–1990s, while VMT growth seems to have slowed to a halt and, since 2007, actually decreased.[20] This shift away from automobile usage and toward public transportation indicates a stark and often unrecognized reversal of trends in travel behavior that have been occurring in the United States since the 1950s.

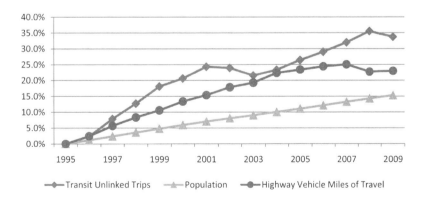

Figure 6.3   Since 1995, transit use has grown more than population or highway travel. (Source: APTA *2010 Public Transportation Fact Book*, 61st ed. [American Public Transportation Association, Washington, DC, April 2010].)

Gasoline Prices to Public Transportation (to Sustainability)

Clearly, one of our enabling factors for oil dependence in the transportation sector is the low cost of driving. The low cost of gasoline, especially relative to the cost in other countries, further emphasizes this in the public consciousness. Table 6.1 displays recent price-per-gallon information for retail fuel prices for a select set of countries compared with the United States.

Cost has been underutilized as a tool to manage automotive travel, particularly in the United States, due partly to the political popularity of the car and partly to the notion that the inelasticity of travel behavior limits the potential of cost to influence driving patterns.[21] Some cost levers such as tolls and high-occupancy lanes have been employed to better incentivize driving decisions in a few cities and to collect additional transport revenues in between cities, but overall the role of pricing in regulating travel behavior has been very limited in the United States. Increasing gasoline prices represent one of the only test cases for the ability of alternative modes of travel, such as transit, to absorb automobile travel that is displaced by increased driving costs.

Other behavioral changes are likely to precede a modal shift in response to gasoline-price escalation. These include increasing personal transportation budgets, reducing VMT through trip chaining and reducing nonessential trips, adjustments to the built environment, and relocation of workplaces or households. Also, increasing automobile efficiency from increasing Corporate Average Fuel Economy (CAFE) standards could decrease even further the elasticity of driving behavior in response to increases in fuel costs.[22] However, some research indicates that driving behavior is

Table 6.1 International gasoline prices, February–March 2012[a]

Country	Price per gallon	Country	Price per gallon
Norway	$9.77	Japan	$7.57
Netherlands	$8.78	Spain	$6.85
Italy	$8.71	Australia	$5.83
Greece	$8.59	China	$5.41
Sweden	$8.37	India	$5.30
Great Britain	$8.03	Canada	$5.26
France	$7.95	United States	$3.52
Germany	$7.72	Mexico	$3.03

[a] Most of these data come from information published in *The Economist*, available at www.economist.com/blogs/graphicdetail/2012/03/daily-chart–18, accessed June 25, 2012. Data were converted from liters to gallons. Data on gasoline prices from China, India, and Mexico are cited from Kshitij Consulting Services, available at www.kshitij.com/research/petrol.shtml, accessed June 25, 2012.

actually becoming more elastic and sensitive to cost levers.[23] Constraints on automotive travel from increasing travel costs due to higher gasoline prices should increase the demand for transit, but the limited deployment of service may mean that the role of transit in alleviating oil dependence is small or altogether irrelevant.

However, despite the lack of attention and utilization in the United States, evidence past and current suggests there is a clear and potentially strong role for public transportation to absorb displaced automobile travel and to alleviate oil dependence. The research on the effect of gasoline prices on travel has rarely focused on public transit, and studies that have examined its relationship to fuel prices have done so almost exclusively in the context of shocks to the pricing system, whether it is the two jolts induced by OPEC to the crude oil market in the 1970s or the spike in prices related to the disruption of access to gasoline after the landfall of Hurricane Katrina in 2005.[24] Relatively little of the already small body of literature attempts to establish some kind of normal or baseline relationship between gasoline prices and transit.[25]

Research generally finds a significant effect that is described in econometric terms as *inelastic*, which means a 1 percent change in gasoline prices results in a less than 1 percent change in transit ridership. Most of the elasticities run from between 0.1 to 0.6, which means that for every 10 percent change in gasoline prices, transit ridership increases by between 1 percent and 6 percent.[26] To illustrate the impact of the elasticities estimated by previous research in the context of their effect on actual numbers of riders, table 6.2 displays the impact of an increase in gasoline prices from $3.00 to $3.75 (an increase of 25 percent) on a city that had an average of 50 million unlinked passenger trips per month at $3.00 per gallon. Such a city would rank 25th in the United States in total yearly unlinked passenger trips in the year 2009.[27] Changes in gasoline prices, as experienced with regularity over the past decade, can lead to millions more or fewer trips taken on transit per year. Though described as significant but inelastic, the effect on transit operations can be quite large, enough those agencies are required to make adjustments in response to changes in cost and ridership.

The literature indicates that trips displaced from automobile to transit are usually peak-period work trips that have replaced short-distance commuting by automobile, though they are also sometimes long-distance commutes where transit is available, such as on a rail line or express bus line. High gasoline prices tend to have a greater impact among travelers who are relatively low-income, central-city residents, and have the effect of moving this population into "transit captivity." This means that, due to the expense of private transport, they were already living at or near an economic level that would limit their travel options to public transit, and increasing gasoline prices have the effect of rendering their automobiles too expensive to drive. Such a process

Table 6.2 Illustration of impact of gasoline prices on ridership in an example scenario: how an increase in gasoline prices from $3.00 to $3.75 affects yearly ridership on a system with 50,000,000 trips per year[a]

Research	Study area and time	Elasticity	Increase in yearly ridership	Percentage change in yearly ridership
Navin (1974)	Midwestern US cities, oil shocks	0.42	5,250,000	10.5%
Agathe and Billings (1978)	Tucson, oil shocks	0.42	5,250,000	10.5%
Wolff and Clark (1982)	Fort Worth, oil shocks	0.26	3,250,000	6.5%
Wang and Skinner (1984)	US cities, oil shocks	0.08	1,000,000	2.0%
		0.80	10,000,000	20.0%
Rose (1986)	Chicago, oil shocks and after	0.11	1,375,000	2.8%
		0.18	2,250,000	4.5%
Mayasuki and Allen (1986)	Philadelphia, oil shocks and after	0.11	1,375,000	2.8%
		0.18	2,250,000	4.5%
Kyte et al. (1988)	Portland, OR, oil shocks and after	0.18	2,250,000	4.5%
		0.30	3,750,000	7.5%
Doti and Adibi (1991)	Orange County, oil shocks and after	0.31	3,875,000	7.8%
Storchmann (2001)	Germany, post-2000	0.07	875,000	1.8%
Currie and Phung (2007)	US nationally, 1998–2005	0.04	500,000	1.0%
		0.38	4,750,000	9.5%
Haire and Machemehl (2007)	US cities, 1999–2006	0.05	625,000	1.3%
		0.54	6,750,000	13.5%
Taylor et al. (2009)	US cities, 2000	0.73	9,125,000	18.3%
		1.45	18,125,000	36.3%
Haire and Machemehl (2010)	US cities, 2002–2007	0.07	875,000	1.8%
		0.10	1,250,000	2.5%
Lane (2012)	US cities, 2002–2008	0.10	1,250,000	2.5%
		0.58	7,250,000	14.5%
Chen et al. (2011)	New York City, 1996–2009	0.12	1,500,000	3.0%
		0.14	1,750,000	3.5%
Stover and Bae (2011)	Cities in Washington State, 2004–2008	0.09	1,125,000	2.3%
		0.47	5,875,000	11.8%
		0.17	2,125,000	4.3%

[a] Adapted from B. W. Lane, "Modeling Urban Transit Ridership in Repeated Cross Sections: What Is the Role of Gasoline Prices in Transit Demand?" (report, MPA Working Paper Series, WP 05–2012, University of Texas at El Paso, 2012), academics.utep.edu/Default.aspx?tabid=71565.

occurs for this population much more rapidly than it does for wealthier travelers, whose additional income gives them a flexibility to choose one of the other options discussed previously (such as improving trip efficiency, changing vehicles, or simply spending more of the personal budget on transport). Additionally, some people who were living near transit may then begin choosing to take it when gasoline prices increase, despite their having incomes well above the economic level of transit captivity. For this population, the convenience costs of a modal shift to transit are relatively low compared with those living farther from access to public transportation.[28]

As gasoline prices go up, this should make driving more expensive, which should encourage reductions in driving and increases in transit patronage. Since most cities, except for a few in the Northeast and on the West Coast, lack the amounts of transit service coverage necessary to serve their populations with access comparable to the automobile, one might expect that only these cities would see any kind of increases in ridership when gasoline prices escalate. However, sprawling, auto-dependent cities in the Sun Belt, the Midwest, and the Mountain West appear to have the greatest elasticities to fluctuating gasoline prices.[29] Table 6.3 shows the largest cumulative elasticities (elasticities added together over multiple monthly lags within a year) from a time-series analysis, and also the largest single elasticities from a cross-sectional analysis to illustrate the cities that appear to have the largest response on transit ridership to gasoline price increases in the United States.

Several important components of the relationship between gasoline prices and transit usage bear further discussion. One is the inelasticity of the response; specifically, the temporal component of any response to fluctuating gasoline prices. The inelasticity of travel behavior to gasoline costs indicates that any adjustment will not be immediate, and the number of other options available for many travelers likewise means that a modal shift to transit may be spread out over time. Research that has studied the temporal nature of the relationship reveals that significant increases in transit ridership will occur several months after a price fluctuation, and continue in seasonal cycles as well. The range in these elasticities is large and volatile across cities and over time.[30] This means that predicting the exact response of transit ridership to price fluctuations can be quite difficult.

Another important component has to do with the interrelationship of gasoline prices to other economic factors. Most of the previous research has focused on gasoline prices and a few proxies for transit service deployment as predictors of transit ridership. However, gasoline prices are but one kind of economic indicator that may influence the cost profile of transport and the decisions motivating travel behavior. More-recent studies indicate that a certain combination of economic characteristics,

Table 6.3 Largest elasticities to gasoline prices[a]

Time-series analysis cumulative lags

Bus city	Lag	Rail city	Lag
Albuquerque	0.77	Seattle	1.12
Cheyenne	0.68	St. Louis	1.12
Kansas City	0.66	Memphis	0.83
Atlanta	0.63	Denver	0.51
Omaha	0.63	Boston	0.48
Indianapolis	0.62	Portland	0.47
Des Moines	0.59	Chicago	0.45
Portland, OR	0.51	San Diego	0.42
Dallas	0.42	Minneapolis	0.39
St. Louis	0.38	Houston	0.35

Cross-sectional analysis largest lags

Bus city	Lag	Rail city	Lag
Indianapolis	0.33	Denver	0.26
Birmingham	0.24	Memphis	0.26
Cheyenne	0.22	Houston	0.18
Houston	0.22	Salt Lake City	0.18
Little Rock	0.20	Dallas	0.15
Atlanta	0.19	Los Angeles	0.15
Milwaukee	0.19	Portland, OR	0.15
Seattle	0.19	St. Louis	0.15
St. Louis	0.18	Boston	0.11
Kansas City	0.16	Miami	0.11

[a] Time series results come from B. W. Lane, "A Time-Series Analysis of Gasoline Prices and Public Transportation in US Metropolitan Areas, Special Section on Rail Transit Systems and High Speed Rail," *Journal of Transport Geography* 22 (May 2012): 221–35. Cross-sectional results come from B. W. Lane, "Modeling Urban Transit Ridership in Repeated Cross Sections: What Is the Role of Gasoline Prices in Transit Demand?" (report, MPA Working Paper Series, WP 05–2012, University of Texas at El Paso, 2012), academics.utep.edu/Default .aspx?tabid=71565.

including gasoline prices, are frequently effective and consistent predictors of transit ridership. These characteristics include rising gasoline prices; a growing number of people in a city's labor force; rising unemployment; and lower automobile costs. The labor-force and employment factors have the strongest effect in cities that have a tight—and tightening—job market; these factors contribute to the increase in the proportion of the population that is transit captive, as noted earlier. This is reflected in the declining automobile expenditures. These factors can work together in seasonal cycles to create economic conditions that encourage increased usage of mass transit. Estimating these factors tends to reduce the wide variance in gasoline price elasticities noted in earlier research.[31]

Additionally, not all transit modes feature the same ridership responses to gasoline price fluctuations. If these factors occur in a growing city with rail transit, the modal shift tends to be stronger on rail than on bus operations. If a city does not have a rail system or the growth of the labor force in the city is stagnant, then a significant modal shift still occurs on bus transit. Rail transit has been noted elsewhere to offer permanence in infrastructure that provides greater certainty about station location, route direction, and timeliness than most conventional bus operations.[32] Rail's higher quality of service and its longer trip length may contribute to its greater attractiveness over bus transit as an alternative mode in response to increasing driving costs. In growing cities, rail services are picking up new residents who have not chosen to move to the suburbs, but are instead seeking out newer transit-oriented neighborhoods that have been a part of the revitalization of transit and central cities for some time now.[33]

It might be expected that a revival of transit demand for the first time in over half a century would represent a boon for transit operators. However, pricing structures dictate that this is not the case, largely because of diesel costs for bus operations. Diesel fuel prices have fluctuated in similar amounts to gasoline prices, and rose nearly 300 percent between 2002 and 2008.[34] Most transit agencies in the United States rely heavily (if not exclusively) on diesel-fueled buses. This means that rising prices increase their operating costs as well. While the increase in ridership increases the farebox recovery ratio, this alone does not mean greater profits for transit. The fact that most ridership increases occur during peak-period travel means an increase in demand when transit operations are already at or near capacity. Additional buses are often needed to meet the increased demand, meaning that transit operations are forced to scale back or eliminate operations on less profitable lines.[35]

How Public Transport Can Help Alleviate Oil Dependence

As discussed earlier, increasing gasoline prices as part of a specific set of economic conditions including a tightening urban labor market and decreasing expenditures available for automobile use has helped increase the demand for transit after decades of decline. Ridership has increased on most modes and in most cities in the United States. However, many agencies have reduced service provisions in order to adhere to their budgets, as ridership is not increasing revenues enough at the fare box to keep up with increased expenses for rising fuel costs to operate service. This is unfortunate for transit's ability to serve the population. One would think that the list of industries and businesses that have been successful in expanding their market share in the midst of increasing demand for their product by raising the cost of the product and reducing its availability is quite short. A focus on meeting short-term budgetary demands at the expense of enhancing service quality and provision can only be bad for the image of transit in the automobile-oriented United States when the price of fueling those automobiles changes to encourage demand for other modes such as public transit.

US cities once boasted public transit systems that were among the best in the world and provided near ubiquitous access within most cities. However, changing attitudes and preferences for automotive ownership and travel, major federal investment and policies facilitating highway construction and less-dense, single-family housing, the advancement of the oil and automotive industry in the United States, and the collusion of oil companies and automakers to purchase rail transit operators and convert those services into bus operations have all helped drive the once-thriving private transit industry into extinction.[36] The role of governmental investment and policy here is particularly intriguing, given the massive federal investment that kept the financial sector afloat after the Great Recession that preceded the writing of this book. One cannot help but wonder how things might have been different for the landscape of transport in the United States and for our role in oil dependence if, 60 or 70 years ago, massive bailouts had been laid out for our nation's major transit companies. Giving them time to innovate to their new economic surroundings, instead of letting the industry die and throwing government support behind a new and burgeoning automobile industry.

The three biggest hindrances to public transportation are low population and employment densities, and a lack of service provision. For mass transit to play a significant and effective role in alleviating oil dependence there must be constraints on automobile use, and transit operations need to be given political freedom and capital to grow their place in the transportation and land-use market. Drivers of policies to alleviate oil dependence include energy security, environmental concerns, and urban redevelopment; they also include financial shortcomings in transportation expenditures and revenues. Transportation, like most contemporary public services, faces serious revenue shortfalls, in the form of the dwindling reserves of the Federal Highway Trust Fund. The fund is financed by the federal gasoline tax, which has remained constant at 18.5 cents per gallon since its inception. Increasing vehicle fuel efficiency, coupled with a flattening of vehicle-miles traveled, means that transportation revenues are not increasing at the pace demanded by new infrastructure construction and deferred maintenance. The result is insolvency of transportation funding in the United States, which has become a political issue and has necessitated several temporary infusions of cash from the general fund in order to maintain funding of transportation projects.[37]

One way or another, revenues have to increase. Altering the federal gasoline tax has thus far been politically unfeasible, but high-profile issues like international economic competitiveness, the priority of infrastructure as a political platform, and threats to safety such as the collapse of the I–35 Bridge in Minneapolis in 2007 can bring enough attention and concern to change the political calculations of what are considered feasible solutions. Additional options for increasing revenues are also possible, such as a taxing mileage traveled, carbon emitted, or other measurements of vehicle usage that better reflect negative externalities associated with driving. However, the gasoline tax does the most to highlight the closest and most immediate connection between the individual users and national oil dependence.

This funding crisis actually presents an opportunity to change the funding structure of public transportation. Transit operators are constrained by tight budgets and are given little room to expand or grow their service. Those that can expand are usually successful at increasing ridership, but this has only been through improvements in service.[38] Public transit struggles to be viable in areas with low densities of population and employment, circumstances that transit has little room to address, since few transit agencies are involved in determining land use and development. If transit agencies could be given the capital and the leeway either to invest in land-use development or to form public-private partnerships to facilitate such development, they could provide the built environment needed to enhance ridership as well as an additional funding source in the form of revenues from their land-use development. Commercial and residential development space in denser, mixed-use urban form typically commands more per square foot to rent or buy than single-use, low-density development. This suggests that potential revenues from investing in this type of development would be significant. Oil dependence can be further alleviated by support for finishing the work already started by transit agencies to decouple public transportation from oil, a process that is much further along for public transportation than the rest of the vehicle fleet. Bio-fuels, natural gas, and hybrids already populate the bus fleet, and the continued electrification of buses plus expansion of intra-city light rail should have negligible impacts on electricity demand. The response to fluctuating gasoline prices indicates that increasing the gasoline tax, in addition to increasing transit ridership on its own, could be used to further finance public transit investments that help alleviate the transportation sector's oil dependence.

Concluding Thoughts on the Case for Transit in Alleviating Oil Dependence

The decline of public transit in the United States was neither inevitable, nor was the rise of the automobile or dependence upon petroleum for energy in transportation. Furthermore, these trends are not necessarily perpetual. Worldwide demand for gasoline is going up, while supplies of petroleum are finite, dwindling, and not located in the most accessible of places.[39] These mathematical certainties dictate that there is no question *whether* we will alleviate our oil dependence. Our energy portfolio is going to undergo a drastic reshaping in the twenty-first century and beyond, which is likely to have major social, economic, and geopolitical consequences. The question is *what* our adaptation away from oil dependence will look like.

Nor should this chapter suggest that a reversion to transit is inevitable, or even the best choice, to alleviate oil dependence in transportation. There are significant changes that will have to be made for transit to have the same utility in the United States as it does in Europe, Japan, and other places that are similarly reliant on public transport for travel. As mentioned earlier, travelers tend to respond to pressures brought by rising gasoline prices in many different ways before they consider a modal shift. Many other solutions—quite a few of which are mentioned in this book—discuss opportunities and potentials for other ways to alleviate oil dependence in transportation that might have greater short- and medium-term utility. In addition to increasing CAFE standards, there are improvements to the internal combustion engine and to structural materials to make vehicles more lightweight, which will contribute to the alleviation of oil dependence in US transportation.[40] There are also significant technological advancements to other powertrains for cars that indicate we may not be driving with only gasoline for much longer anyway. Virtually every automaker in the world has an electric or hybrid model either available now or due to arrive in the near future, and almost every country in the world similarly has goals to replace up to 15 percent of their domestic fleets with electrics or plug-in hybrid cars within the next decade. Hydrogen and the fuel cell have also seen extensive research and investment in their potential as automotive fuels.[41]

In any case, public transit as a solution to oil dependence offers other societal benefits that modifications in automotive technology and driving behavior alone, which support existing development patterns, fail to provide. There is significant evidence that built environments that facilitate walking, biking, and the use of public transit encourage healthier lifestyles, while urban sprawl and auto-dependence further facilitate a sedentary lifestyle that contributes to the US obesity epidemic.[42] Removing automobiles can contribute to relieving traffic congestion, and a more densely built environment served by transit limits the ecological footprint of urban development that is otherwise seen in auto-oriented urban form.[43] Land development that is denser and more mixed-use than traditional suburban sprawl development continually prices itself at a premium in the land-use market, suggesting that there is a latent demand that has not been met for the way of living provided by an environment that is conducive to transit usage (see chapter 13 by John Renne).

Gasoline prices are only going to increase in the long term, and we will transition away from gasoline usage as we know it in transport. Early market signals indicate that, though there are relatively few efforts to support such a trend, people will move trips over to transit when the price of gasoline increases. Public transportation has a useful and potentially necessary role in alleviating oil dependence in transportation in the United States, and for it to achieve its potential there must be the kind of significant policy support that facilitated oil dependence in transportation in the first place.

Notes

1. Energy Information Administration, "Weekly US Regular Conventional Retail Gasoline Prices" (EIA report, 2011), www.eia.doe.gov/petroleum/data_publications/wrgp/mogas_history.html, accessed June 25, 2012.

2. Energy Information Administration, "Demand for Oil and Gasoline" (EIA report, 2011), www.eia.doe.gov/pub/oil_gas/petroleum/analysis_publications/oil_market_basics/demand_text.htm, accessed February 8, 2011.

3. Bureau of Transportation Statistics, *National Transportation Statistics 2010* (BTS publication, Research and Innovative Technology Administration, Washington, DC, 2010), www.bts.gov/publications/national_transportation_statistics/.

4. For example, see: W. R. Black, *Transportation: A Geographical Analysis* (New York: Guilford Press, 2003).

5. W. R. Black, *Sustainable Transportation: Problems and Solutions* (New York: Guilford Press, 2010). Black details the main issues in US transportation that render it unsustainable. These include the high levels of automobile use in the United States, which have several negative externalities contributing to the lack of sustainability in current transportation systems. Petroleum is nonrenewable and finite, and while worldwide demand is increasing, remaining supplies are located in either geologically or geopolitically difficult places to access. A primary

by-product of burning petroleum is carbon dioxide, which is the primary greenhouse gas facilitating global climate change. Additionally, burning petroleum releases a host of other polluting gases that contribute to urban air pollution. Congestion exacts real economic and health costs on us, while the automobile and its roadway networks kill over 40,000 people per year, in addition to contributing to obesity and other social problems.

6. For seminal coverage of the relationship between land use and travel, see: M. G. Boarnet and R. Crane, *Travel by Design: The Influence of Urban Form on Travel* (New York: Oxford University Press, 2001).

7. Black, *Transportation*.

8. D. L. Greene, "Transportation and Energy," chap. 10 in S. Hanson and G. Guiliano, *The Geography of Urban Transportation*, 3rd ed. (New York: Guilford Press, 2004), 274–93. David Greene has published extensively on the use of energy in transportation. See also: *Transportation Energy Data*; the most recent version is available at cta.ornl.gov/data/index.shtml.

9. Indiana University, *Plug-in Electric Vehicles: A Practical Plan for Progress* (publication of the School of Public and Environmental Affairs, Indiana University, Bloomington, IN, February 2011), www.indiana.edu/~spea/pubs/TEP_combined.pdf.

10. Federal Transit Administration, *Public Transportation's Role in Responding to Climate Change* (FTA publication, US Department of Transportation, Washington, DC, January 2009), www.fta.dot.gov/documents/PublicTransportationsRoleInRespondingToClimateChange.pdf, accessed June 26, 2012.

11. See: 2009 National Household Travel Survey, table 9: "Annual Number (in Millions) and Percent of Person Trips by Mode of Transportation and Trip Purpose," and table 26: "Usual Mode to Work vs. Actual Commute Mode on Travel Day" (Federal Highway Administration, US Department of Transportation, 2009), nhts.ornl.gov/publications.shtml.

12. Ibid., table 9; see also: 2011 National Transportation Statistics, table 1–40, "US Passenger Miles" (report, Research and Innovative Technology Administration, Bureau of Transportation Statistics, US Department of Transportation, 2011), www.bts.gov/publications/national_transportation_statistics/2011/html/table_01_40.html.

13. American Public Transportation Association, *2011 Public Transportation Fact Book*, 62nd edition (APTA publication, Washington, DC, April 2011). Several metropolitan areas in the Sunbelt, Midwest, and Western Mountain regions of the United States have made significant investments in enhancing public transport service and infrastructure, and the impact of these investments has been notable. However, they still represent only a blip on the screen of total transit usage and operations in the United States.

14. Black, *Transportation*.

15. Ibid.

16. Ibid.

17. American Public Transportation Association, *2010 Public Transportation Fact Book*, 61st edition (APTA publication, Washington, DC, April 2010).

18. C. Hass-Klau, G. Crampton, M. Weidauer, and V. Deutsch, *Bus or Light Rail: Making the Right Choice*, 2nd ed. (Brighton, UK: Environmental and Transport Planning, 2003).

19. For example, see: D. Gordon, *Steering a New Course: Transportation, Energy, and the Environment* (Washington, DC: Island Press, 1991); see also: P. O. Muller, "Transportation and Urban Form: Stages in the Spatial Evolution of the American Metropolis," chap. 3 in Hanson and Guiliano, *The Geography of Urban Transportation*, 59–85. The commercial and emotional connections to home ownership and automobile use have been well documented in many places and do not merit further discussion here.

20. American Public Transportation Association, *2011 Public Transportation Fact Book*.

21. The literature on driving elasticity is reviewed in: T. Litman, *Changing Vehicle Travel Price Sensitivities: The Rebounding Rebound Effect* (publication of the Victoria Transport Policy Institute, Victoria, BC, Canada, 2012), www.vtpi.org/VMT_Elasticities.pdf, accessed June 25, 2012.

22. US Corporate Average Fuel Economy (CAFE) standards are scheduled to increase from 27.5 mpg for passenger vehicles in 2010 to 54.5 mpg for model year 2025. See NHTSA press release, www.nhtsa.gov/About+NHTSA/Press+Releases/2011/President+Obama+Announces+Historic+54.5+mpg+Fuel+Efficiency+Standard, accessed June 25, 2012.

23. Litman, *Changing Vehicle Travel Price Sensitivities*. A report from the US Congressional Budget Office also suggests this; see: *Effects of Gasoline Prices on Driving Behavior and Vehicle Markets* (publication of the Congressional Budget Office, Washington, DC, 2008).

24. Research on the OPEC oil shocks and their effect on public transit include the following:

D. E. Agathe and R. B. Billings, "The Impact of Gasoline Prices on Urban Bus Ridership," *Annals of Regional Science* 12, no. 1 (1978): 90–96.

American Public Transportation Association, "Impact of Rising Fuel Costs on Transit Services: Survey Results" (APTA publication, Washington, DC, May 2008), www.apta.com, accessed January 21, 2009.

J. Horowitz, "Modeling Traveler Response to Alternative Gasoline Allocation Plans," *Transportation Research* 18A, no. 2 (1982): 117–33.

D. L. Keyes, "Energy for Travel: The Influence of Urban Development Patterns," *Transportation Research* 16A, no. 1 (1982): 65–70.

D. Masayuki and W. B. Allen, "A Time-Series Analysis of Monthly Ridership for an Urban Rail Rapid Transit Line," *Transportation* 13, no. 3 (1986): 257–69.

F. P. Navin, "Urban Transit Ridership in an Energy Supply Shortage," *Transportation Research* 8 (1974): 317–27.

M. C. Nizlek and L. Duckstein, "A System Model for Predicting the Effect of Energy Resources on Urban Modal Split," *Transportation Research* 8 (1974): 329–34.

J. Sagner, "The Impact of the Energy Crisis on American Cities Based on Dispersion of Employment, Utilization of Transit, and Car Pooling," *Transportation Research* 8 (1974): 307–16.

G. H. K. Wang and D. Skinner, "The Impact of Fare and Gasoline Price Changes on Monthly Transit Ridership: Empirical Evidence from Seven US Transit Authorities," *Transportation Research* B 18, no. 1 (1984): 29–41.

G. J. Wolff and D. M. Clark, "Impact of Gasoline Price on Transit Ridership in Fort Worth, Texas," *Transportation Engineering Journal* 108, no. 4 (1982): 362–75.

Research prompted by or related to the spikes seen around Hurricane Katrina include the following:

G. Currie and J. Phung, "Transit Ridership, Auto Gas Prices, and World Events: New Drivers of Change?" *Transportation Research Record* 1992 (2007): 3–10.

G. Currie and J. Phung, "Understanding Links between Transit Ridership and Gasoline Prices: Evidence from the United States and Australia," *Transportation Research Record* 2063 (2008): 133–42.

A. R. Haire and R. B. Machemehl, "Impact of Rising Fuel Prices on US Transit Ridership," *Transportation Research Record* 1992 (2007): 11–19.

A. R. Haire and R. B. Machemehl, "Regional and Modal Variability in Effects of Gasoline Prices on US Transit Ridership," *Transportation Research Record* 2144 (2010): 20–27.

B. W. Lane, *An Exploratory Analysis of the Influence of Gasoline Prices on Public Transit: Evidence from US Metropolitan Areas* (Unpublished PhD dissertation, Indiana University, July 2010).

B. W. Lane, "Effect of Fuel Price Increases on US Light Rail Ridership 2002–2008: Results and Implications for Transit Operation and Policy," *Transportation Research Circular* E-C145, 63–77 (publication of the Transportation Research Board, National Research Council, Washington, DC, 2010).

B. W. Lane, "The Relationship between Recent Fuel Price Fluctuations and Transit Usage in Major US Cities," *Journal of Transport Geography* 18, no. 2 (2010): 214–25.

V. W. Stover and C.-H. C. Baue, "Impact of Gasoline Prices on Transit Ridership in Washington State," *Transportation Research Record: Journal of the Transportation Research Board* 2217 (Washington, DC: Transportation Research Board of the National Academies, 2011), 11–18.

25. Research that focuses more on the baseline relationship between gasoline prices and public transit includes the following:

C. Chen, D. Varley, and J. Chen, "What Affects Transit Ridership? A Dynamic Analysis Involving Multiple Factors, Lags, and Asymmetric Behaviour," *Urban Studies* 48, no. 9 (2011): 1893–1908.

B. W. Lane, "A Time-Series Analysis of Gasoline Prices and Public Transportation in US Metropolitan Areas—Special Section on Rail Transit Systems and High Speed Rail," *Journal of Transport Geography* 22 (May 2012): 221–35.

B. W. Lane, "Modeling Urban Transit Ridership in Repeated Cross Sections: What Is the Role of Gasoline Prices in Transit Demand?" (report, MPA Working Paper Series, WP 05–2012, The University of Texas at El Paso, 2012), academics.utep.edu/Default.aspx?tabid=71565.

G. Rose, "Transit Passenger Response: Short- and Long-Term Elasticities Using Time-Series Analysis," *Transportation* 13, no. 2 (1986): 131–44.

K.-H. Storchmann, "The Impact of Gasoline Taxes on Public Transport: An Empirical Assessment for Germany," *Transport Policy* 8, no. 1 (2001): 19–28.

26. See table 1 in: Lane, "Modeling Urban Transit Ridership in Repeated Cross Sections."

27. APTA, *2011 Public Transportation Fact Book*.

28. The information contained in this paragraph is summarized from the literature citied in notes 24 and 25 above.

29. These results come from the following:

Lane, "An Exploratory Analysis of the Influence of Gasoline Prices on Public Transit."

Lane, "The Relationship between Recent Fuel Price Fluctuations and Transit Usage in Major US Cities."

Lane, "A Time-Series Analysis of Gasoline Prices and Public Transportation in US Metropolitan Areas."

Lane, "Modeling Urban Transit Ridership."

30. Chen et al., "What Affects Transit Ridership?"; see also: Lane, "A Time-Series Analysis of Gasoline Prices and Public Transportation in US Metropolitan Areas."

31. Lane, "An Exploratory Analysis of the Influence of Gasoline Prices on Public Transit"; and Lane, "Modeling Urban Transit Ridership."

32. Hass-Klau et al., *Bus or Light Rail*.

33. Lane, "An Exploratory Analysis of the Influence of Gasoline Prices on Public Transit"; and Lane, "Modeling Urban Transit Ridership"; and Lane, "Modeling Urban Transit Ridership." The intent of transit-oriented development has been researched and discussed at length elsewhere; research of its impact on travel behavior includes: A. J. Khattak and D. Rodriguez, "Travel Behavior in Neo-Traditional Neighborhood Developments: A Case Study in USA," *Transportation Research Part A: Policy and Practice* 39, no. 6 (2005): 481–500.

34. American Public Transportation Association, "Impact of Rising Fuel Costs on Transit Services: Survey Results" (APTA publication, Washington, DC, May 2008), www.apta.com, accessed January 21, 2009.

35. Ibid.; and Storchmann, "The Impact of Gasoline Taxes on Public Transport."

36. Gordon, *Steering a New Course*; and Muller, "Transportation and Urban Form," 59–85.

37. Congressional Budget Office, "The Budget and Economic Outlook: Fiscal years 2011 to 2021" (CBO publication, Washington, DC, January 2011).

38. Hass-Klau et al., *Bus or Light Rail*.

39. International Energy Agency, *Key World Energy Statistics* (Paris: IEA, 2011), www.iea.org/textbase/nppdf/free/2011/key_world_energy_stats.pdf.

40. Indiana University, *Plug-in Electric Vehicles*.

41. Ibid.

42. Lawrence Frank is among those who have studied this problem extensively; see, for example: H. Frumkin, L. D. Frank, and R. Jackson, *Urban Sprawl and Public Health: Designing, Planning, and Building for Healthy Communities* (Washington, DC: Island Press, 2004).

43. R. Ewing, R. Pendall, and D. Chen, *Measuring Sprawl and Its Impact* (publication of Smart Growth for America, Baltimore, MD, 2000), www.smartgrowthamerica.org.

7 Taking the Car out of Carbon

Mass Transit and Emission Avoidance

PROJJAL K. DUTTA

The transportation sector is the largest consumer of fossil fuels and the largest emitter of greenhouse gases (GHG) in the United States. Top-billed strategies to reduce fuel consumption, namely "fleet fuel-efficiency" and "decarbonization" are not likely to deliver the required reductions of fuel consumption and GHG emissions. For the transportation sector to meaningfully reduce its footprint, there has to be a shift away from personal automobiles to public-transportation—and the lifestyle that comes with it. Livable communities where walking and bicycling are viable, and where residences and businesses are smaller and situated more closely together, are the solution, rather than a continuation of the auto-centric paradigm, even with greater engine efficiency and/or electrification of fleet. Conceptually, the big reduction opportunity lies not in taking the carbon out of the car, but in taking the car out of carbon. In other words creating communities where the car is not the only means of transportation or even the preferred means; rather than communities where cars still rule, with a modification that powers them with electricity.

A recognition of public transportation's role in preventing carbon emissions, followed by quantitative evaluation, and finally, a pricing of the environmental benefits, can result in a means to fund this fundamental shift in transportation, and therefore in fossil fuel consumption.

How Big Is the Transportation Slice?

It is well understood that transportation is responsible for the bulk of fossil-fuel consumption in the United States. What is less well understood is that transportation is also responsible for a very large—close to 43 percent and consequently the single largest share of all GHG emissions. In fact, the translation from fuel to emissions, essentially amplifies the impact of transportation. This amplification is not very well recognized, even in professional literature. A commonly cited Environmental Protection Agency (EPA) statistic states that about 32 percent of all American GHG emissions come from the transportation sector.[1] This is a number that reports tailpipe emissions only, thereby substantially understating the true impact of transportation. Approximately 50 percent of what is emitted at the tailpipe is also emitted, as a prequel, in the refining of the fuel. Even without accounting for the embodied energy, and the attendant emission, that goes into the manufacture of vehicles and construction of transportation infrastructure (e.g., bridges, highways, etc.); and counting only the emissions at the tailpipe plus emissions from refining and transporting the fuel ("wells-to-wheels"), transportation-related emissions are 43 percent of all US emissions, the single largest slice of the entire emissions pie.[2]

Further, as easily extracted fossil sources of petroleum dry up, the newer supplies tend to be those that take a greater amount of energy to refine from raw material into usable fuel. Oil sands in Canada's Alberta province, as compared to crude pumped out of the Saudi Arabian desert, present an excellent case in point. What that means is that the GHG emissions prequel of fossil fuels is actually increasing rather than decreasing. So while today it adds about 50 percent to the current tailpipe emissions, in the future it could possibly add more, thereby likely increasing the total impact of transportation-related emissions.

It is critical to make a distinction between this emissions prequel to fossil-fuel use and "embodied" energy. The "wells-to-wheels" fuel calculation is inherently different from the embodied energy that goes into the construction of the highway or the automobile. Embodied energy is amortized over the life of the structure or the car. The wells-to-wheels energy, on the other hand, has a direct relationship with the fuel that is combusted in the engine; it cannot be divided over a large denominator if, say, the car lasts an additional year. Thus, these emissions can and should be grouped together with the tailpipe emissions. Unlike those of the EPA, there are other standards, including one from the World Resources Institute, that use the wells-to-wheels metric.

Transportation-related energy consumption is also an excellent index for overall energy use / fossil-fuel consumption and GHG emissions of a society. While it is true that developed societies consume more energy on a per-capita basis than developing societies, levels of development and the per-capita energy consumption do not share a linear correlation. It is possible, indeed quite common, for countries with higher development indices—infant mortality, literacy, longevity, etc.—to have lower per-capita energy and fuel consumption than societies lower on those same indices. In other words, there is a point of inflexion beyond which per-capita energy and fossil-fuel consumption levels can increase without a commensurate increase in living standards. This point of inflexion is a key threshold between societies that have good public transportation and those that do not.

As illustration, the transportation-related GHG emission percentage for Germany is 28 percent.[3] That is to say, transportation accounts for 28 percent of all GHG emissions in Germany, adding both tailpipe and the refining-related emissions; the equivalent number for the United States is 43 percent. The Germans not only expend a smaller fraction of their entire energy and emissions on transportation but, in large measure due to their lower transportation-related emissions fraction, also consume much lower amounts of energy / fossil-fuels and, not surprisingly, have lower GHG emissions per capita as compared to the Americans—this despite the fact that Germans live longer, have a vibrant industrial sector manufacturing great cars among other products, have no speed limits on the highways, and, surprisingly, have higher rates of auto-ownership than Americans. Yet they do all this by consuming less fuel and emitting about half the GHG as compared to Americans on a per-capita basis. (Germans tend to be on the higher side of the per-capita emissions in the non-American/Canadian industrialized world. Similar numbers—i.e., per-capita fossil-fuel consumption, energy expenditure, and GHG emissions from the United Kingdom or Japan are lower than even those of Germany.) The underlying factor in all of these examples is the presence and high utilization of public-transportation in the case of Germany (or the United Kingdom or Japan) and the absence and relatively little utilization of public transportation in the case of the United States.

The vast difference in fossil-fuel / energy usage, especially the energy used in getting around, between the average American and the average German arises not from what goes on in outlying communities or on the farm, but rather from the fact that most metropolitan areas in the United States are essentially a sea of auto-dependent sprawl, while those in Germany and large parts of Europe have viable and well-used mass transit. The few American cities, such as New York, San Francisco, Chicago, or Boston, that have good public transportation systems and accompanying density have per-capita fuel / energy consumption and GHG emission that are not only lower than other parts of the United States, but in some cases lower than those of Europe.[4]

Rebound Effect: The Supply-Side Problem

"Decarbonization," a strategy or suite of strategies currently billed as the ticket from here to there, that is from our fuel-consuming, emissions-spewing present to a carbon-free future, addresses only the supply side of the problem. Any sustainable solution will have to address both supply *and* demand sides of the equation.

Generating energy renewably with little or no GHG emissions is chief among the currently fashionable strategies, with wind turbines and solar photovoltaic (PVs) being the poster children. These are, undoubtedly, a large part of the solution. However, to think that these technologies and others that are still in development will be able to supply all the energy that we consume, without our having to reduce any of our current energy consumption, is unrealistic. The trend lines, very troublingly and not surprisingly, point in the exact opposite direction; the increase in the appetite to consume has and continues to outpace the gains made through clean energy supply.[5] Renewable energy as a fraction of total energy production in the United States has hovered around the 1 percent mark for the past several years. This 1 percent includes biomass, mostly the burning of woodchips and other forest by-products. If total renewable energy were to be calculated without biomass, it's about half of 1 percent of total energy production. As David Owen, writing recently in the *New Yorker*, commented wryly, the percentage of renewable energy production in the United States, rounded off to the closest whole number, is zero.[6] In that context, to be dreaming of a world where energy is first generated renewably and then used to power automobiles, is exactly that that—dreaming.

Making commodities "cheaper," whether through reducing their cost or through more efficient use of them, usually makes the demand for them go up, not down. Specifically in the world of transportation, this phenomenon has some bearing upon fleet fuel efficiency, which is being positioned as the primary means of reducing GHG emissions. In the United States, several state-, regional-, and metropolitan-level Climate Action Plans (CAPs) seek to achieve a significant portion of their GHG emissions reductions through increased fleet efficiencies. Simply put, the hope in these plans is that statewide (or regional) fuel combustion and, therefore, emissions will go down because cars will start getting more miles to the gallon. There is, unfortunately, little historical evidence to support such a hope. What history shows to be more likely is that, as cars become more efficient, people will drive them more. In the past couple of decades, cars have indeed become more fuel-efficient. The gains made through efficiency, though, have been more than wiped out by increased vehicle-miles traveled (VMT) and an even greater increase in VMT per capita. Recent flattening of VMT trends are encouraging, but coming as they do in the wake of historic unbridled growth, they are still at unsustainable levels. The joke—if you believe that transportation planners indulge in humor—is that that VMT is tapering off only because one person cannot drive two cars at the same time. But then again, Google is doing everything it can to change that.[7]

One way to think of fuel efficiency is in terms of the cost of driving the car per mile.

If mileage per gallon doubles (with, hypothetically, the cost of the vehicle remaining the same), then the immediate impact is that the cost-per-mile becomes half of what it used to be. In cost terms that same end could be achieved if the cost of gas were halved. In other words, the doubling of fuel efficiency and the halving of the price of gas have an identical cost impact—they both halve the cost per mile. So, regionally across the United States, it should follow that in places where the gas is cheapest (or, in efficiency terms, where engines are most efficient), car owners would spend the least on getting around. As a fraction of household income, though, the reverse is true. It is precisely in the regions where gas is the cheapest that the largest fraction of income is spent on driving. And as a corollary, it is where gas is the most expensive (or the engine is the least efficient) that the smallest fraction of income is spent on driving.[8] There is in this data a possible reversal of the efficiency case—rather than efficiency, engine *in*efficiency may hold the answer!

Even if that were an exaggeration, there is not enough evidence to show that increasing fuel efficiency is going to result in a commensurate reduction of fuel consumption or GHG emissions. Fuel efficiency will produce some gains, but within the larger context of growing size of vehicles and the increasing trip lengths, a 50 percent efficiency gain will have a much smaller than 50 percent fuel-consumption / GHG emission reduction. Interestingly, average fleet efficiency today is about the same as that of the Ford Model T. The intervening hundred years have not seen a discernible trend toward overall efficiency. And when there have been across-the-fleet fuel-efficiency enhancements, it has typically been in response to a very strong price signal. Once that signal has faded away, efficiency of the fleet has gone down.[9] To have this same fleet fuel efficiency as the central strategy of climate action planning seems to only serve the purpose of delaying the inevitable—wasting more precious time while GHGs in the Earth's atmosphere increase—before coming to the conclusion that it's not more-efficient driving but a minimization of driving that is key to helping solve the global-warming problem from a transportation standpoint. To reiterate, it is the car that has to be taken out of the carbon, not the carbon out of the car.

Good Fuel-Consumption (Good Carbon)

Traditional discussion about sustainability and mass transit are usually focused on technologies that consume less fuel or electrical power, facilities that have solar panels on the roof, and rolling stock that regenerates energy. The focus, in other words, is on the means and methods that reduce the fuel consumption and/or carbon footprint of mass transit. As valuable as such discussions are, this thinking misses the main point. Mass transit's footprint—in terms both of its GHG emissions and also its own fossil-fuel consumption—has an inverse relationship to the regional, and therefore global, carbon footprint. As long as transit's carbon footprint is not increasing due to inefficiencies, regional GHG emissions and fuel consumption will go down if transit's footprint *increases* rather than decreases. The increased fuel consumption—being a predictor for increased service, which, in turn, is a predictor for increased ridership—is, therefore, a measure of reduced dependence on the less fuel-/carbon-efficient personal automobile. Transit operations in Europe, for instance, consume more fuel than transit operations in the United States. This only makes the entire transportation sector there more efficient, not less.

A helpful concept for understanding this notion—that transit's increased fuel use is a reduction in overall fuel use—is "carbon prevention." An analogous situation can be seen in the manufacture and deployment of a wind turbine. There is, undoubtedly, a certain amount of energy consumption / greenhouse-gas (GHG) emissions in the manufacturing process, transportation, and assembly of a wind turbine. However, the energy that it generates (and the fuel it eventually helps save or the GHG emissions that it helps avoid) far outstrips that which goes into its manufacture and deployment. In other words, while it is critical for the power-generation industry as a whole to cut its fossil-fuel consumption and its GHG emissions, it is actually desirable, indeed critical, for the wind-turbine manufacturing segment of that industry to increase its fuel consumption and GHG emissions. The latter's proportionately small increase signals the former's much larger subsequent reduction. We can think of such emissions as "Good Carbon"—that is, emissions that cause a consequent and larger reduction of emissions. For overall emissions to reduce, Good Carbon emissions have to go up. In other words, while we should be aiming for a set of strategies that require reductions across the board, it is much more nuanced and actually beneficial for the world, and indeed for overall reduction of dependence on fossil fuels and GHG concentrations, for some sectors to increase their consumption and emissions.

Mass transit and several other sectors of the economy, renewable electricity generation included, have to play a more prominent role in a world where fossil-fuel consumption / GHG emissions are on a decreasing curve. This important role will come with an increased GHG footprint as well. As long as each unit of increased fuel consumption/emissions is accompanied by an even larger (hopefully many times larger) unit of consumption/emission avoidance, then this increase is not something to be prevented; rather it is something to be embraced.

The most dramatic illustration of "carbon avoidance" in the United States comes from the transportation sector. Fuel consumption and, consequently, emissions from an automobile passenger-mile (one mile traveled by one person—the typical mode for

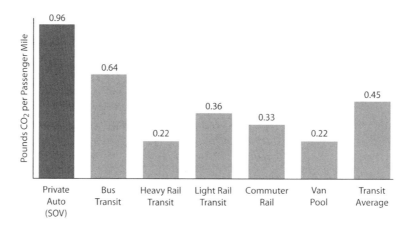

Figure 7.1   Estimated CO_2 emissions per passenger-mile for transit and private autos. (Source: Sightline Institute, www.sightline.org.)

most Americans) are several times more than the equivalent from a transit passenger-mile (fig. 7.1) and, of course, from bicycling and walking, which are activities that are synergistic with transit. A subway car or a bus consumes a certain amount of energy to maintain and run. We can think of this as its debit side. On its credit side, it removes its passengers from automobiles, generates denser land-use (which in turn has significant fuel-consumption-reduction / emission-reduction benefits), makes for communities where people can walk and ride bicycles, and it reduces traffic, allowing the remaining automobiles to operate with improved efficiency.[10] In light of each of these benefits, the transit passenger-mile generates a very high rate of carbon-avoidance. Studies show that, through these benefits, the carbon-dioxide emitted by transit helps avoid many times as much as would be emitted from private automobiles. A comparison of energy consumption data for various regions in the United States clearly illustrates this net saving. Residents of New York City, on average, consume a fraction of the energy that the average American consumes. In fossil-fuel terms, the relative advantage of transit-served communities is even greater.

Transit Effect Multiplier: Fuel for Transit Operations Is Recouped Many Times Over, as Savings

Mass transit's role in fossil-fuel usage reduction has been intuitively and qualitatively understood for a long time. However, there has been scant academic or industry-wide attention devoted to quantifying this phenomenon—in terms of emission. The American Public Transportation Association (APTA) set up a Climate Change Working

Group (CCWG) in 2007 under the aegis of its Sustainability and Urban Design Standards (SUDS) Committee. This working group drew together representatives from more than a dozen operating agencies, including those from Seattle, Portland, San Francisco, Los Angeles, Chicago, Washington, and New York State's Metropolitan Transportation Authority. Members also included professionals from consulting firms, nonprofits, and academia. Together this group, via meetings, conference calls, and draft after draft, came up with a methodology for measuring transit's GHG emission impacts.

The APTA methodology essentially broke up transit's GHG emissions into two categories. The first included GHG emitted directly or indirectly by a transit operation, which could therefore be thought of as the "debit" side, and the second category comprised GHG emissions avoided in the region served by the transit operation—the "credit" side (fig. 7.2).

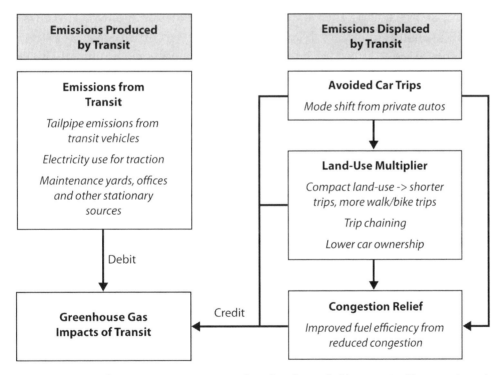

Figure 7.2   Greenhouse-gas emissions produced and avoided by transit. (Source: American Public Transportation Association, "Quantifying Greenhouse Gas Emissions from Transit" [APTA report, SUDS-CC-RP-001-09, 2009].)

Each of the displacement components, in turn, can be quite nuanced. For instance, Mode Shift, as a factor, varies from one part of the MTA's service territory to another. In Midtown Manhattan, for instance, the mode options may be walking or bicycling, which are less fossil-fuel / carbon-emission intense than mass transit, while on the edges of Long Island, the only mode option would be an automobile, which is far more carbon-intense than public transit. Consequently, the overall regional Mode Shift factor has to be a weighted average. Likewise, the Land-Use Multiplier, which accounts for things like the trip-not-taken or reduced trip-length as a result of sustainable density engendered by the presence of transit, is high for Midtown Manhattan but low for suburban Long Island. Congestion Relief reflects the fact that road networks function more efficiently (crawling traffic lessens the efficiency of car engines), when there are parallel transit networks reducing the load.

The MTA, as a part of its Blue Ribbon Commission on Sustainability, engaged the consulting firm Booz Allen Hamilton to study the environmental effects of its operations. One of the primary mandates for Booz was to apply the APTA methodology to quantify GHG emissions avoided by the MTA. The consultants utilized survey data, spatial analysis, and other statistical methodologies to model the MTA network's carbon-reduction impact. It concluded that the carbon avoidance, or as Booz termed it, the "Transit Effect Multiplier"—a composite of Mode Shift, Land-Use Factor, and Congestion Relief, amounted, on average, to 8.24. In other words, for every unit of GHG that the MTA emits, it helps New York City and the rest of the MTA's 5,000-square-mile region avoid about eight and a quarter times that in carbon emissions. As governments at the national, regional, and local level take steps to reduce GHGs, each operator in the transit sector around the world will need to calculate and verify the GHGs they emit and, more important, the GHGs they avoid.

The MTA also signed up to become one of the founding members of The Climate Registry in the spring of 2008. This nonprofit organization provides the framework for reporting GHG emissions for organizations in the public and private sector, according to a uniform, publicly available and transparent methodology laid out in their General Reporting Protocol. The Registry, in effect, is like the Lands Records Office of a new economy. They record the carbon "inventory" (a word they prefer) of various organizations, both public and private, and keep a record of this inventory in the public domain. The Registry also requires that members contract with an approved independent third-party to verify their annual emissions report. The MTA successfully reported and verified its 2008 GHG emissions to TCR as 2.3 million metric tons carbon equivalent.

Putting these two numbers together gives us a sense of the net carbon benefits of the MTA network. Multiplying 2.3 million metric tons times 8.24 (Transit Effect Multiplier / Carbon Avoidance Factor), we get 18.95 million metric tons. If we subtract from this the 2.3 million metric tons that the MTA emitted, we get 16.65 million metric tons. In other words, setting aside what the MTA emits, it is still responsible for avoiding approximately 17 million metric tons of GHG emissions throughout the region. At the rate of $30 per ton (a price projection in the Waxman-Markey climate bill, passed by the US House of Representatives), the MTA's carbon avoidance would be worth about $500 million.

While these estimates are likely to change as related protocols and standards change, it still becomes abundantly clear from this exercise that the MTA and other transportation agencies provide a very valuable carbon-reduction service, which would hold equally true in terms of fossil-fuel usage reduction, which is currently not recognized, priced, or monetized.

Selling Prevented Emissions

Across large swathes of the economy today there is a value placed on delivering a good or service in a "green" way. As utility customers we can all purchase electricity or, for a premium, we can purchase "green" electricity. While this market is quite nascent and entirely voluntary in the United States at the retail level, it is nonetheless a real monetary recognition of the "green" attribute and value of power-generation. Likewise, many automobile manufacturers will now offer the same car in a regular versus a hybrid model. These cars, identical in most of their physical characteristics, carry a substantial price-tag difference—a difference not explained by the reduced fuel consumption alone. There is, in that price difference, recognition and a pricing of an additional value that the hybrid car brings to its owners as also to the entire world.

Unlike hybrid cars or renewably generated electricity, transit has not been successful in attributing and monetizing its environmental benefit. That is, commuters and governments alike pay for / subsidize transit operations for their ability to move riders from A to B—not, critically, for the fact that this moving from A to B happens in a fuel-efficient and carbon-efficient way. The fact that transit provides this movement in a way that produces fewer emissions per passenger-mile traveled does not currently feature in the pricing. This pricing or valuation does not have to translate into additional fare. In fact, the ideal situation would be the reverse. As long as there is a standardized protocol and established methodology for quantifying and verifying this avoidance, it can be commoditized such that the transit agency generating the avoidance is able to sell that on the carbon markets. This would be similar to the wind-power generator being able to sell the energy that the turbine generates and additionally being able to sell the "green" attribute of that energy, sometimes called the Renewable Energy Credit or REC. Both the energy and the REC have a measurable value and both of those values are generated within a market-based system. A protocol that allows public transportation to sell its carbon avoidance would set up a virtuous cycle in which the incoming resources could be used to rebate fares, to increase the extent of their networks, and so on. In effect, an old-fashioned fossil-fuel-burning power plant or gas-guzzling automobile will underwrite some fraction of the cost of transit, making it more affordable to more people—hopefully, thereby, increasing ridership and further cutting GHG emissions within a region.

At the current moment, carbon markets mandated by law ("mandatory carbon markets") exist only in Western Europe and in Japan. In the absence of a mandatory market, "voluntary carbon markets" do emerge—as they have in the United States. However, with no mandated cap on emissions, the value of a unit of carbon emission or prevention never rises to a level where it can influence the financial decisions made by corporations or jurisdictions. In mandatory markets, though, that is not the case, and a unit of carbon trades at levels (at the time of writing, about $30 per metric ton) that end up influencing choices of equipment, projects, phasing, and so on.

As the world moves toward a long-overdue system of measuring and limiting its GHG emissions, there is an opportunity for the transit sector to make its case as a key, measurable, and verifiable contributor to helping slow climate change and reducing overall oil usage, which in turn should enable transit agencies to seek higher levels of investment in transit modes and networks. Transit's benefits accrue not only in the developed world, but also in fast-developing economies such as India and China, where transportation funds are being competitively allocated between the automobile and mass transit, often in favor of the former. In order to educate policy makers of the carbon-reduction benefits of investing in transit, the sector has to come together, build a robust, rigorous case, and then spread the word. The MTA's carbon-avoidance factor, a product of the research by the MTA and APTA, presents exactly that opportunity.

Co-Benefits

The presence of good public transportation is key to a community's, a region's, and a nation's reduced dependence on fossil fuels and consequent emissions of GHG. However, even if we did not care about fossil-fuel consumption and GHG emissions, we could still make the argument that there are a larger set of benefits for which there should be increased investment in transit—the so-called co-benefits. Co-benefits are non-transportation-related secondary benefits that accrue to places where there is an underpinning of viable and well-used public transportation.

Good public transportation is a key component of the public realm; it is like a town square where millionaires and the masses rub shoulders—because it provides them both the quickest means of getting from one place to another. It engenders communities with sustainable densities. Development coalesces around transit lines; there is nothing like steel rails on the ground as a market signal. People move their households, they seek work and recreation along the route, and a livable community is formed where once there was sprawl. This has been historically true. Low-density communities through which subway lines were built ended up being livable, walkable places in New York, Boston, and other cities; and very interestingly, it continues to work today in erstwhile sprawl, such as along the new alignments in Los Angeles. Here, slowly but steadily, development has sought out transit lines, and densities have gone up and continue to go up.

The usual cost-benefit analyses, which look at the price of every ton of GHG emission prevented or every gallon of fuel saved, use small discounting periods of only 15–20 years for evaluating infrastructure projects that often take most or all of that time period for realization. Consequently, building transit capacity shows up in these analyses as an expensive way to mitigate GHG or save fossil fuels. However, if one accounts for the slow, gradual nature of development and extends the discounting period to account for cross-generational trends, there are few strategies as effective, or for that matter as inexpensive, as transit to counter GHG emissions and reduce fossil-fuel dependence. Evidence on the ground, such as household-emissions comparisons between regions with good transit and those without, illustrates this clearly. Consider the per-capita energy consumption comparison, earlier in this chapter, between New York City and the American average (see note 4); that comparison, essentially, holds between many other similar communities with good public transportation and communities without.

The presence of viable public transportation tends to generate communities that are much denser than those that are largely auto-dependent. These denser communities usually have physically smaller homes (and businesses) that are located closer to each other, or in planning terms, a higher ratio of "dwellings per acre." While in places like Manhattan that number is about 150 units per acre, parts of Brooklyn and Queens (or for that matter Cambridge, Massachusetts; Palo Alto, California; or much of urban Europe) the density tends to be around 50 dwellings per acre. At these densities, anything above 10 dwellings per acre allows for and encourages a lifestyle that includes "nonmotorized" trips, such as walking and bicycling. Urban densities also engender the "trip not taken." For instance, it is possible to pick up groceries from the store on the way from the transit stop to home, thus making a separate trip to the grocery store unnecessary. This same density that is at the heart of successful and widespread bicycle and pedestrian strategies also makes the average motorized trip length much shorter. Denser communities, therefore, reduce another key planning metric—vehicle-miles traveled (VMT).

It is well understood that denser communities are good for the environment. Smaller homes and businesses require less heating, air-conditioning, electricity, and water, and they take less in materials and energy to construct and less energy to maintain. What is not understood as well, however, is that such densities are good not only for the air and the water around us but also for ourselves. Research undertaken by a research team at Rutgers University establishes an inverse correlation between national obesity rates in developed Western nations and their rates of public transportation use coupled with biking and walking.[11] The United States, with its lowest use of the combination of walking and biking, also has the highest obesity rate. This trend is discernible within the granularity of the United States, too.[12] Washington, DC, with its highest use of public-transportation, biking, and pedestrian trips, has the third-lowest obesity rate in the nation. Similarly, states like New York, Massachusetts, Connecticut, California, and Oregon show a very high inverse relationship between their national ranking in transit or "bikeped" (broken out into bicycling and pedestrian separately) usage and their obesity ranking. This relationship (with the exception of Colorado, where outdoor lifestyle choices become countervailing factors) is true in reverse as well. Mississippi, Tennessee, and Alabama have among the lowest transit and bikeped usage and the highest obesity rates.

Conclusion

Public transportation is, arguably, the best and most effective strategy that the United States has for reducing its consumption of fossil fuels, and in turn for reducing its dependence on foreign oil imports and reducing its GHG emissions. Transit is also the most overlooked strategy in all of these respects. It is overlooked for many reasons, some of which are structural, such as the lack of a powerful industry to lobby on its behalf. Others, like 20-year discounting models for initiatives that take more than 20 years for realization, could be addressed relatively easily. So although analysis based on current methodologies shows public transportation to be not hugely efficient or effective as a means of reducing oil consumption and GHG emissions, the vast weight of empirical evidence is in its favor. On a per-capita or per-household basis, the least fuel-consumptive jurisdictions in the United States—Washington, DC, New York City, San Francisco, Boston, etc.—are all served by good public transportation. This holds true not just across space, but also over time. In other words, communities that have built or extensively refurbished and extended their public transportation systems in recent memory—such as Los Angeles and Portland, Oregon—have seen rising commuter numbers, increased urban density, and, consequently, reduced fuel use.

However, public transportation has taken the backseat (no pun intended) in groupings of strategies to combat global warming and oil dependence. This is not an accurate reflection of its potential. It is a proven technology with verifiably astounding results, and it has additional co-benefits in the fields of public health, energy, and national security as well as the ability to stimulate economic growth in regions where it provides service. It offers people transportation options that, in many instances, they currently do not have. Where people do have a public transportation option, they choose it in overwhelming numbers. It is a democratizing force that lifts the waters for entire regions in addition to providing benefits for the entire planet.

Sepia-toned pictures of an America from a hundred years ago show bustling urban communities in the exact places that have today been completely ravaged by planning for the automobile. In these pictures, fit-looking men and women walk on the streets, with streetcars running behind them. While we may not want to go back to those days entirely, there is much to be learned from them. Our generation will have to stop patching up what the people in those pictures constructed by way of public transportation infrastructure. We will have to invest significantly to revitalize the infrastructure that is still standing and we will have to invest additionally to create new infrastructure for the next century. Such modern, efficient, world-class infrastructure will generate economic activity and will result in our children and grandchildren being able to design their communities and their lives around simple activities like walking back from the grocery store with bags in their hands, or riding bicycles to work, or taking modern, efficient, and convenient train rides to the places where they live, work, and play.

Notes

1. See: table ES–7: "US Greenhouse Gas Emissions Allocated to Economic Sectors (Tg or million metric tons CO_2 Eq.)," *Inventory of US Greenhouse Gas Emissions and Sinks: 1990–2010.*

2. Federal Transit Administration, "Public Transportation's Role in Responding to Climate Change" (report, FTA, US Department of Transportation, Washington, DC, 2010), www.fta .dot.gov/documents/PublicTransportationsRoleInRespondingToClimateChange2010.pdf.

3. World Resources Institute, "Energy and Resources Country Profiles, Germany" report published by EarthTrends, The Environmental Information Portal, earthtrends.wri.org/pdf _library/country_profiles/ene_cou_276.pdf.

4. US Energy Information Administration, "Independent Statistics & Analysis" (country-by-country listings of total primary energy consumption, energy intensity per capita, and total primary energy consumption; EIA report, US Department of Energy, Washington, DC, 2012), www.eia.gov/emeu/international/energyconsumption.html; City of New York, *Plan NYC Progress Report 2012* (publication of the City of New York, 2012), nytelecom.vo.llnwd.net/o15 /agencies/planyc2030/pdf/PlaNYC_Progress_Report_2012_Web.pdf. In this report, New York City's per-capita energy consumption is calculated to be 146.9 MBTU.

5. US Energy Information Administration, *Annual Energy Review: Primary Energy Production by Source, 1949–2010*, table 1.2 (EIA publication, US Department of Energy, Washington, DC, October 19, 2011), www.eia.gov/totalenergy/data/annual/showtext.cfm?t=ptb0102.

6. David Owen, "The Artificial Leaf: Daniel Nocera's Vision for Sustainable Energy," *The New Yorker*, May 14, 2012.

7. "Google Cars Drive Themselves, in Traffic," *New York Times*, October 9, 2010, www.ny times.com/2010/10/10/science/10google.html?pagewanted=all.

8. "The Varying Impact of Gas Prices"—Interactive feature. *New York Times*, June 9, 2008, www.nytimes.com/interactive/2008/06/09/business/20080609_GAS_GRAPHIC.html#.

9. Kate Pickert, "A Brief History of Fuel Efficiency," *Time*, January 29, 2009.

10. American Public Transportation Association, "Quantifying Greenhouse-Gas Emissions from Transit" (report, American Public Transportation Association Standard APTA SUDS-CC-RP-001-09, August 14, 2009), www.aptastandards.com/Portals/0/SUDS/SUDSPublished /APTA_Climate_Change_Final_new.pdf.

11. David R. Bassett Jr., John Pucher, Ralph Buehler, Dixie L. Thompson, and Scott E. Crouter, "Walking, Cycling, and Obesity Rates in Europe, North America, and Australia," *Journal of Physical Activity and Health* 5 (2008): 795–814.

12. See, for example: awesome.good.is/transparency/web/1008/driving-and-obesity–3 /flat.html.

8 High-Speed Rail and Reducing Oil Dependence

PETRA TODOROVICH AND EDWARD BURGESS

In 2008–2009, the US Congress passed a series of bills authorizing and appropriating funds to improve passenger rail service in the United States. The bills included provisions for building new high-speed rail systems of the sort that can only be found in Europe and Asia today. These funds added up to $10.1 billion in grants for passenger rail, of which $3.9 billion have been devoted to building a new, statewide, high-speed rail system in California. In total, 32 states have received planning or construction grants for passenger rail under the new High-Speed Intercity Passenger Rail Program launched by the Obama administration in 2009 (fig. 8.1).

However, the commitment of the US Congress to this new program has been fickle. The 112th Congress charted a different course from that of the 111th, which had supported high-speed rail, providing no new funding for the high-speed and passenger rail program and recapturing funds left unspent by three new Republican governors who rejected projects that had been committed to by their predecessors.[1] Yet, despite the increased politicization of the program on Capitol Hill, and with the exception of those governors, the program has enjoyed wide support. Thirty-nine states sought and applied for funding from the new high-speed rail program, proposing projects with costs adding up to $75 billion. Rail ridership in the United States has grown as well. Amtrak posted its most robust year of ridership in 2012, carrying 31.2 million passengers—the highest number of passengers since it started providing service in 1971 (fig. 8.2).[2]

The recent funding commitment by the federal government, participation in its rail program by 32 states, and growing passenger rail ridership volumes all suggest

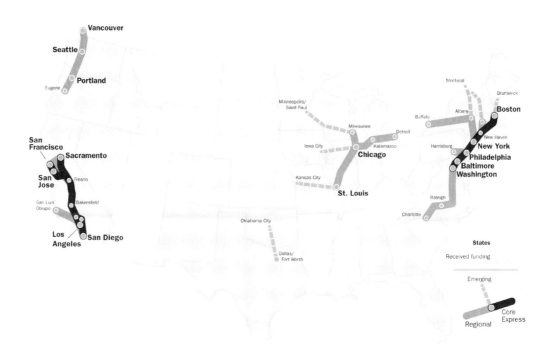

Figure 8.1   Summary of investments made through High-Speed Intercity Passenger Rail Program as of October, 2011. (Source: US Department of Transportation, 2009.)

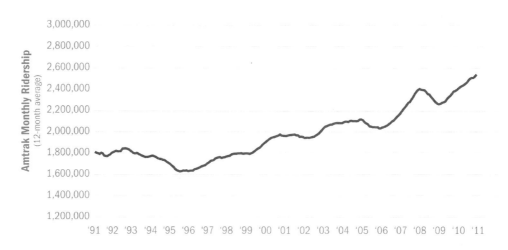

Figure 8.2   Amtrak ridership has demonstrated a steady increase over the last two decades. (Source: US Department of Transportation, Bureau of Transportation Statistics, 2011.)

that interest in expanding passenger rail in America is likely to outlast the current political opposition at the congressional level. Accordingly, the role of passenger rail and, in particular, high-speed, electrified rail is examined in this chapter for the role it can play in contributing to reducing America's dependence on oil and environment-threatening air pollution.

Oil Dependence, Air Pollution, and Intercity Travel

The passenger transportation sector is unmistakably the largest component of US oil consumption (fig. 8.3). Additionally, passenger transportation energy provided by fossil-fuel combustion (primarily of petroleum products) contributes significantly to local and regional air pollution as well as global greenhouse-gas emissions (fig. 8.4).

Thus, determining which forms of transportation infrastructure can deliver passenger travel while minimizing energy consumption is a key problem for planners, engineers, and policy makers seeking to reduce oil dependence and mitigate environmental impacts. In particular, transportation infrastructure projects with long lifetimes (such as airports, highways, and railroads) should be given careful consideration

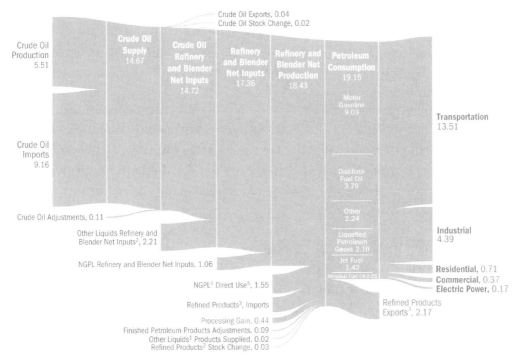

Figure 8.3   Petroleum flow, 2010 (millions of barrels per day). (Source: US Energy Information Administration, 2010.)

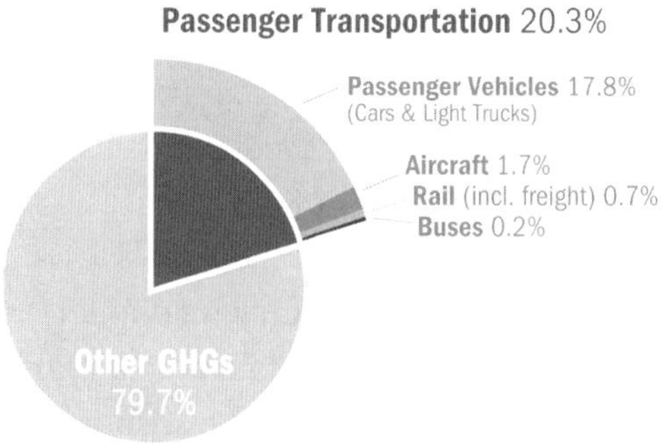

Passenger Transportation 20.3%

Passenger Vehicles 17.8%
(Cars & Light Trucks)

Aircraft 1.7%
Rail (incl. freight) 0.7%
Buses 0.2%

Other GHGs
79.7%

Figure 8.4 Passenger transportation
contributes a large share of the US greenhouse-
gas emissions. The majority of these emissions
are from petroleum-product combustion in
private automobiles. (Source: US Environmental
Protection Agency, 2011.)

before their construction, since these capital investments could alter day-to-day transportation choices for decades or longer.

The vast majority of passenger trips in the United States consists of daily (i.e., local) travel. However, intercity travel makes up a disproportionate share of total miles traveled, due to the longer average distance per trip. Recent survey estimates suggest that long-distance trips make up about a third of the *total distance* traveled by passengers in the United States, while the intercity fraction of *total trips* is much lower.[3]

From an energy perspective, reducing the negative impacts of long-distance intercity travel poses a unique challenge, since options for alternative travel modes are limited. Many policies have been proposed for reducing the impact of transportation by targeting modes that currently supply passenger travel. These policies aim to lower energy consumption and carbon emissions via improvements in vehicle technologies and fuels. Fewer policies focus on addressing the demand side of the transportation sector, by shifting passenger travel demand toward lower impact through reduction in trip length, more efficient land-use, and an emphasis on public transportation. One exception to this is the public support for high-speed rail, which has been promoted by the US Department of Transportation as a means to address both oil dependence and greenhouse-gas emissions.[4]

In this context, this chapter considers the potential for high-speed rail as a means of reducing emissions and oil consumption from US passenger travel. We demonstrate that high-speed rail has theoretical advantages, in terms of direct energy utilization, over other possible transportation modes. However, the extent of this advantage and the subsequent potential for high-speed rail to have a direct impact on oil consumption and emissions is highly dependent on a variety of other factors, and may be much smaller than some proponents suggest. Moreover, we comment on prospects for high-speed rail to bolster a more efficient transportation network and provide other benefits.

Overall Market Potential for High-Speed Rail

The potential for high-speed rail to directly reduce total oil consumption and emissions in the United States is constrained by the fraction of passenger travel demand that this mode is able to accommodate—specifically intercity trips, probably in the 100- to 800-mile range. This excludes most daily trips, which, as previously mentioned, constitute the bulk of US passenger travel (approximately 67 percent of US person-miles traveled), as well as longer-distance trips (approximately 18 percent), such as cross-country airplane flights. Thus, the remaining 15 percent of current person-miles traveled constitutes a reasonable upper bound of passenger travel that high-speed rail might theoretically serve. Scenario analyses can provide more detail into the potential for high-speed rail to address the dual challenges of oil consumption and greenhouse-gas emissions. According to one recent scenario analysis, the potential for CO_2 reductions from large-scale high-speed rail investment is on the order of 1 percent.

These estimates suggest that the direct benefits from high-speed rail in terms of overall energy and emissions may be modest. However, these analyses also neglect the indirect impacts in terms of land-use and city-centering, which may be large and are difficult to measure and attribute. It is our view that these indirect benefits may be more important than any direct reduction in energy utilization as passengers choose high-speed rail over alternative travel modes.

Notwithstanding this limited market potential, the following sections provide a more detailed comparison of high-speed rail and other transportation modes in terms of its direct energy and emissions trade-offs.

Comparing Energy Consumption of Travel Modes

Any attempt to compare the merits of publicly funded transportation infrastructure projects should employ a common framework to evaluate the operational costs and benefits of these projects. Ideally, this framework would consider external social costs such as the project's contribution to environmental damage (i.e., pollution) and oil dependence. This can be a challenge, since these external costs are seldom included in the private cost of transportation service provision. Furthermore, transportation infrastructure is heavily subsidized, and the demand for constructing new roads, railways, and airports may not reflect the true value of each mode if these external costs were included. Since many of these external costs are broadly associated with energy consumption, one possible comparison metric (one that is often used in life-cycle assessment studies) is the amount of energy consumed, e.g., megajoules (MJ) per passenger-distance traveled, for example, passenger-kilometers traveled (PKT).

We will employ this metric (MJ/PKT) to provide a general sense of how transportation investments (particularly high-speed rail) rank in terms of minimizing oil consumption and environmental damages. It must be noted, however, that equivalent distances for passenger modes may not be equivalent in terms of services provided; values such as flexibility, speed, accessibility, and comfort are not considered in the simplistic assessment provided by MJ/PKT. Nonetheless, the authors believe that a comparative approach utilizing MJ/PKT can highlight some key trade-offs for policy makers to consider among possible transportation investments.

Physical Energy Requirements for Intercity Vehicle Operation

At a minimum, the energy inputs needed to power a vehicle must be sufficient to overcome the fundamental physical forces resisting motion, such as friction (rolling resistance), aerodynamic drag, and inertia.[5] The fundamental physical forces necessary for travel are in turn dictated by characteristics such as vehicle weight, speed, drag-coefficients, and so on. As an illustration of this, figure 8.5 shows how energy requirements scale up with increasing speed for a theoretical high-speed train. Despite the fact that energy use increases exponentially with train speed, empirical measurements have suggested that high-speed trains can achieve lower energy use per PKM than regional trains due to greater investment in energy-efficient technologies and operations.[6]

Overall energy consumption in vehicle technologies is subject to not only these physical energy requirements but also inefficiencies occurring during energy conversion from the energy source (e.g., from the electric wire or the gasoline tank) to the engine or motor and ultimately the wheels. As an example of this conversion inefficiency, consider the automobile schematic in figure 8.6.

As the diagram illustrates, engine inefficiencies cause large energy losses resulting in only about 30 percent energy-conversion efficiency from the fuel tank to the drivetrain (driveline).[7] By comparison, electric trains employ highly efficient inductive motor technologies that frequently operate near thermodynamic limits. One estimate

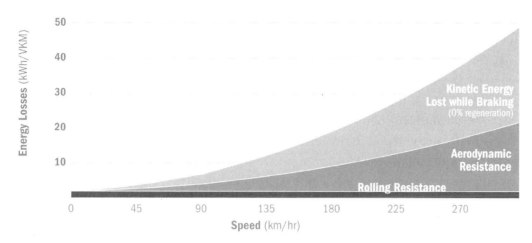

Figure 8.5   Energy losses as a function of train speed. Parameters used to model these values are shown in Burgess, 2011. This chart shows a basic case with no regenerative braking and no correction for changes in speed between stops.

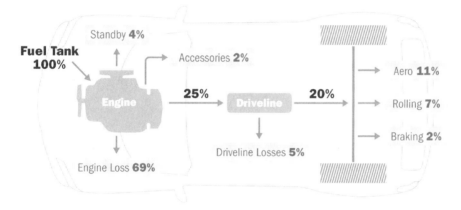

Figure 8.6   Energy losses during an automobile drive cycle. (Source: Transportation Research Board and National Research Council, 2006.)

suggests that power conversion efficiencies on trains are on the order of 85 percent.[8] These estimates suggest that there may be relatively more room for efficiency improvements in automobiles than in trains.

While it is possible to foresee incremental improvements in many vehicle characteristics (such as more-streamlined trains to reduce drag, or lightweight materials to reduce train mass), there are practical limitations to improvements beyond a certain point. For instance, drag cannot be eliminated entirely and will always create a limitation for vehicle energy use, especially for vehicles traveling at high speeds.

These physical realities constrain the degree to which vehicle improvements alone can reduce energy utilization. For high-speed rail, one of the most promising technologies for reduced energy consumption appears to be regenerative braking.[9] Current estimates show that 35–40 percent of kinetic energy normally lost to braking can be recovered through this technology.[10] Even greater energy recovery might be achievable with advances in energy-storage technology.[11]

Assuming incremental improvements toward a thermodynamic minimum energy usage for trains, opportunities to improve energy utilization may still exist via improving the passenger loading of trains. Additionally, the petroleum content of the fuel can be reduced through replacement with biofuels or electrification.

Energy Comparisons of Transportation Modes

Due to varying physical parameters, conversion efficiencies, and life-cycle costs, passenger transportation modes have drastically different energy use per vehicle-kilometer traveled. Public transportation modes with larger, heavier vehicles that create more aerodynamic drag (such as trains and airplanes) unsurprisingly require much larger energy expenditures per miles traveled. However, these differences can be counterbalanced by the ability of heavy vehicles to spread these expenditures out on a per-passenger basis. As such, the energy and emissions performance of many transportation modes are highly dependent on passenger load factors. In one of the most comprehensive comparisons to date, energy use and emissions were evaluated across various transportation modes.[12] High-speed rail was found to vary from <1 MJ/PKT for a very high-occupancy scenario (100 percent seats full) to >7 MJ/PKT for a very low occupancy (10 percent of seats full) scenario. This compares to energy use just below 2 MJ/PKT for automobiles and just above 2 MJ/PKT for airplanes at their average occupancies. This illustrates that high-occupancy trains have the capacity to outperform other transportation modes, but the same trains may underperform if load factors are too low to yield a superior MJ/PKM value.

Technological Progress and Carbon Emissions for Vehicle Operation

While this and other life-cycle studies have made great headway into understanding the full impact of transportation on resources and the environment, it is also important to recognize that these comparisons are not static. By focusing on the underlying vehicle technology, we intend to illustrate that impacts are not set in stone and will change over time as technology improves or ridership patterns change. Vehicle-technology evolution within the lifetime of large infrastructure projects can alter the relative advantages of each transportation mode. For example, private automobiles are currently subject to CAFE standards related to fuel efficiency (MPG), which will change the performance of these vehicles over time (assuming these policies are maintained by future elected and appointed officials).

To grasp the effect of technological change on high-speed rail's energy and emissions performance, it is helpful to compare forecasts of different types of passenger-vehicle technologies. The comparison illustrated in figure 8.7 is based on forecasts for the following five transportation modes (methodology and assumptions for these results are presented in more detail in Burgess's report "Sustainability of Intercity Transportation Infrastructure"[13]):

1. Personal automobiles—fuel economy derived from AEO 2011 CAFE6 scenario for on-road MPG of vehicle fleet.
2. Aircraft—derived from projections in Lee for improvements in airplane fuel efficiency (extrapolated through 2050).[14]
3. High-speed rail with low-carbon energy source—modeled forecast from Burgess,[15] representative of electricity sources in California.
4. High-speed rail with high carbon energy source—modeled forecast from Burgess,[16] representative of electricity sources in the Midwest United States.
5. Battery electric vehicle—based on current technologies and no modeled improvements and electricity sources in California.

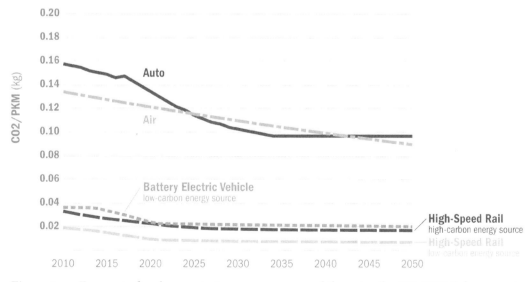

Figure 8.7   Forecast of carbon emissions per passenger-kilometer (kg CO_2/PKM) for several different intercity travel technologies.

The results illustrate that even under scenarios with significant technological improvements, rail modes can potentially outperform other modes. However, as explained previously, the extent of this advantage depends heavily upon the passenger load factor and other operational parameters. The most important operational factors determining the outcome of train performance include load factor, train capacity, maximum speed, and acceleration/deceleration (which, in turn, is affected by the frequency of stops, operating profile, and presence of regenerative braking).

As mentioned above, passenger load factors (i.e., ridership) have an outsized influence over the per-capita performance of transportation modes compared to other factors. Thus, investments in rail routes with low potential for ridership are likely to underperform or even be detrimental in terms of energy use and emissions. Furthermore, these operational estimates exclude life-cycle impacts of high-speed rail and the energy and emissions footprint associated with construction.

Life-Cycle Impacts and Payback Times for Carbon Emissions

So far our discussion of transportation energy consumption has focused exclusively on operational impacts. However, for infrastructure-intensive projects such as high-speed rail lines, highways, and airports, it is paramount that planners include the full life-cycle impacts of these projects. Indeed, no advantage in operational energy consumption can be meaningful if it is outweighed by the up-front energy footprint from construction. Construction materials such as concrete, steel, power systems, and the like embody significant energy and emissions footprints that occur well in advance of any possible reductions. One way of accounting for these life-cycle costs is to spread them over each PKT in the life of the project. (This approach is discussed further in the following section.)

Alternatively, a payback-period assessment is one way to consider these initial inputs and determine how beneficial the investment may be in the long run. A variety of factors can influence the energy payback period of high-speed rail, most notably the ridership of the system.

For instance, one study showed that the proposed California high-speed rail line could have an energy payback time of eight years under a high-ridership scenario (and low ridership among other modes), but under a low-ridership scenario, the construction energy inputs may never achieve payback. Another recent study focusing on carbon emissions shows a similar result, with the payback time for the initial balance of carbon emissions depending most crucially on ridership but also on other factors like vehicle technology and operations (fig. 8.8).[17]

The high-speed rail system initially contributes a net positive contribution to CO_2

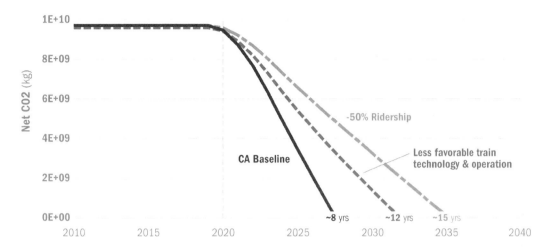

Figure 8.8 Carbon-emissions payback timeline for several modeled high-speed rail scenarios in California.

emissions due to life-cycle emissions embedded in construction materials. After operation commences in 2020 (indicated by vertical dashed line), passengers switching to high-speed rail from other modes leads to a gradual reduction in cumulative emissions relative to business-as-usual today. This is the contribution that can be attributed to the high-speed rail system. The net emissions gradually decrease until achieving a payback several years later. The modeled results shown in the figure above are detailed in Burgess's "Sustainability of Intercity Transportation Infrastructure"[18] and represent the following scenarios: (1) California baseline scenario using assumptions in California high-speed rail planning efforts as of 2010 (solid line); (2) California baseline with less-favorable performance in train technology and operations (dashed line); (3) California baseline with 50 percent fewer riders than current projections (dotted line).

Ridership Diversion and Induced Travel

Another important concept for planners to focus on is that high-speed rail, in isolation, contributes its own footprint to energy consumption and carbon emissions. In theory, this operational footprint is offset by reduction of travel in other modes, thereby providing a net reduction in energy and emissions. However, this reduction may not materialize in the event that high-speed rail lines are primarily occupied by "induced travel," where new trips occur that would not have previously occurred.

Assuming high-speed rail has an advantage over other modes in terms of per-capita energy consumption, then this benefit is only achieved if there is some diversion of travel (either current or future additional travel) from another mode to high-speed rail. In other words, if ridership on high-speed rail is solely in addition to existing travel demand, and the passengers choosing to travel via high-speed rail do not reduce the amount of travel on other modes, then high-speed rail will have no net reduction on oil consumption or environmental impacts.

Some forecasts for high-speed rail (for instance, in the Northeast Corridor, e.g., Amtrak 2010) demonstrate a very high induced-travel demand, suggesting that many riders will represent additional travel growth but will not mitigate current oil consumption or environmental impacts by reducing the number of automobiles on the road. Assuming that high-speed rail has an advantage over other modes in terms of per-capita energy consumption, this benefit is only achieved if there is maximum diversion of travel from alternative modes to high-speed rail.

In a growing economy, some increase in travel is likely to be inevitable, and decisions must be made about what type of infrastructure should be built to accommodate this additional travel. In this context, induced travel on lower-intensity modes such as high-speed rail may be preferable if it provides a *relative* advantage over induced travel on alternative modes (i.e., new highway construction). However, it's important to recognize that the infrastructure may not bring *absolute* reductions in energy and emissions. Thus, we stress the importance of making infrastructure decisions that reflect a fair comparison across modes.

Another advantage offered by high-speed rail is the potential to reduce secondary trips. The spatial requirements of airport runways tend to put them on the outskirts of a city and require additional trips via automobile or public transit (if available). However, high-speed rail has the unique advantage of transporting passengers directly to the city center and closer to likely destination points and transit systems.

Vehicle Electrification and Primary Energy Sources

An important conclusion drawn from the forecasts in figure 8.7 is that high-speed rail has only a slim advantage over private automobiles if battery electric vehicles are widely adopted. This outcome is likely to take some time due to the slow turnover rate of the on-road vehicle fleet.[19] However, it reveals the impact that vehicle electrification has on reducing carbon emissions (and oil consumption), whether this is accomplished via private means (i.e., automobiles) or public (i.e., rail) means.

If reduction in domestic oil consumption is considered to be an important social goal, then the potential for electrification of long-distance travel is a primary advantage that high-speed rail offers over internal-combustion-engine vehicles as well as airplanes. One exception to this advantage is battery electric vehicles (e.g., Nissan Leaf). Currently, battery electric vehicles that are ready for wide-scale market adoption are not capable of traveling the long distances required for intercity travel. However, some technology providers are poised to overcome this obstacle through improvements in battery technology[20] or the addition of battery-swap stations.[21]

In the United States, diesel locomotives currently power most intercity rail routes. However, globally, most high-speed rail systems are electric. While incremental improvements to current diesel-powered systems are likely to continue in the near term, a transition to true high-speed rail in the United States would likely necessitate electrification and bring with it the associated costs and benefits of this transition.

It is important to remember that electrification of intercity travel does not displace energy demand, but merely shifts it to the electric power system. Thus, a complete cost-benefit assessment of high-speed rail or other transportation system must include the resource and environmental impacts of this additional demand on the electric grid.

Since the 1970s, the United States has largely eliminated the use of petroleum as a primary energy source for electric power. Thus, electrification of transportation could go a long way toward reducing oil consumption. However, the overall impact on emissions is less clear-cut. Currently, there are large regional disparities in primary energy sources of electric power in the United States. For instance, the Midwest relies largely on coal-fired power plants, while the Pacific Northwest relies largely on hydroelectric power plants. Thus, a unit of electricity produced in these two places may have very different costs in terms of air pollution and carbon emissions produced. The interconnected nature of the electricity grid and the inability to attribute electrons to their generation source makes the true impacts of electricity difficult to determine at any point in space or time. Even states with strong support for renewable energy often import large amounts of nonrenewable sources, such as coal, from out of state. Recent attempts have been made to take these imports into account and still suggest a strong regional variability in electric emissions intensity.[22] Thus, regions with a lower carbon fuel mix, and strong renewable portfolio standards, may be better targets for high-speed rail deployment if emissions reduction is seen as a desirable policy outcome.

Variations in Market Demand and Energy Use by Region

Regional factors such as population density and land-use patterns may also influence ridership demand for high-speed rail services, which in turn impacts the operational energy efficiency and life-cycle benefits of high-speed in reducing oil use and carbon emissions. A recent study evaluated the relative potential market demand for passenger rail service on existing rail rights-of-way in the United States by evaluating a set of criteria that measured land-use, demographic, and transportation-infrastructure characteristics.[23] The authors weighted factors such as regional populations, concentrations of employment in central business districts, transit connectivity of the population and employment, and congestion on competing modes in order to create an equation that measured potential market demand for high-speed rail on close to 8,000 rail corridors in the United States (fig. 8.9).[24]

This study suggested a framework for evaluating regional differences that may contribute to demand for rail services. These regional differences impacting ridership demand also make a difference in the number of secondary trips associated with the location of a rail station within the region. For instance, for two regions of comparable population size but different land-development patterns, the rail station at the center of the more centralized region served by public transit is likely to result in fewer secondary auto trips than the rail station located in the decentralized region with ample parking and convenient highway access, which would act more like an airport in the number of auto trips it generates. This concept can be illustrated in the comparison of Philadelphia and Houston, below, two regions of roughly six million people each

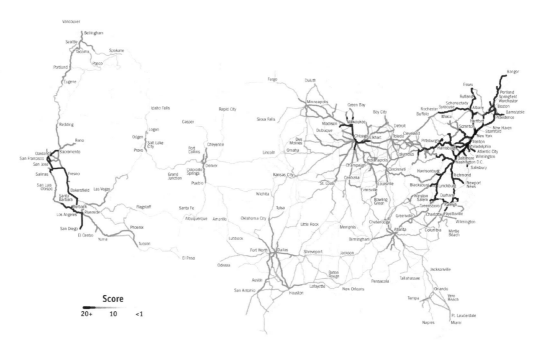

Figure 8.9   Relative market demand of potential passenger rail corridors. (Source: Todorovich and Hagler, 2011.)

but with very different spatial development patterns at their city cores.[25] As shown in figure 8.10, approximately 220,000 people live within two miles of Philadelphia's 30th Street rail station. This contrasts with the center of the Houston's region (lacking a central train station, the authors used the center of the region's downtown as the center point) in which 72,000 people live, roughly one-third of the population around Philadelphia's train station. This difference in population density and transit networks, also shown in figure 8.10, suggests that high-speed rail connecting to Philadelphia's train station would be readily accessible by pedestrian trips, public transit, and short car trips, compared to Houston, where the majority of access would likely be by automobile, no matter where in the region the station may be located. Thus, in much the same way that regional differences in energy supply will impact the carbon footprint of high-speed rail operations, so will regional difference in population density and land-development patterns impact the number of secondary auto trips associated with high-speed rail, and thus its ability to reduce oil demand.

Taking this reasoning one step further, locational decisions of high-speed rail stations within their regional context may also have an impact on the pollution and

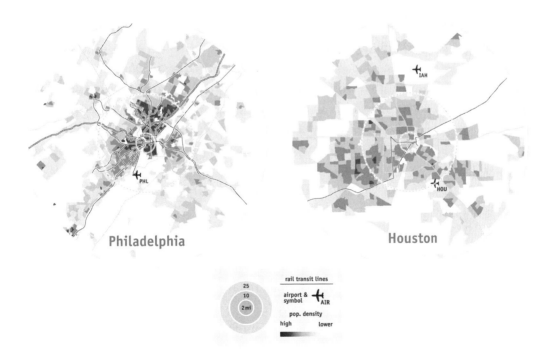

Figure 8.10   Population density and transit networks in Philadelphia and Houston. (Source: Todorovich and Hagler, 2011.)

energy-reducing potential of high-speed rail. There are examples in Europe (Avignon, France, and Camp de Tarragona, Spain, for example) in which high-speed rail stations were located on the edge or periphery of cities or small towns, with ample parking and minimal or no pedestrian or transit access. In addition to generating relatively low ridership, these stations have had little impact on land-use patterns, failing to attract development around the station.[26] Due to the energy and carbon efficiencies of dwellings and transportation patterns in urban areas, we assume that intensified land development around train stations would provide a benefit over stations located in peripheral areas where the potential for compact development that makes use of existing infrastructure is less. High-speed rail can encourage more-compact development patterns around station areas, but as shown below, concerted strategies must be applied to ensure that new stations have an impact on urban redevelopment.

Impact of High-Speed Rail Stations on Urban Revitalization

There is ample evidence in European case studies of high-speed rail stations acting as catalysts for urban revitalization in central city core or edge areas.[27] These benefits are often experienced when a conventional rail system is upgraded to high-speed rail, and the associated improvements to the infrastructure and station, such as tunneling or bridging over rails and rail yards, create new space in urban centers for commercial development and public-realm improvements.[28] It is important to note, however, that these benefits are not due to provision of rail infrastructure alone, but dependent on a set of strategies that contribute to successfully maximizing the impact of high-speed rail on urban revitalization. In a study of six European high-speed rail stations, Ribalayguay and García distill three types of successful strategies in promoting urban revitalization with high-speed-rail interventions: prevision strategies, project strategies, and promotion strategies.[29]

Prevision strategies refer to actions that anticipate the physical needs around the rail station, such as reserving land for commercial development, transportation connections to the center city, and/or zoning and policies to allow development around the train station. *Project strategies* refer to the characteristics of the rail service, such as timetables, frequency, prices, the new station itself, and urban design around the station; and *promotion strategies* refer to actions that promote the rail service, including marketing the rail service as well as tax incentives and policies geared toward tourism and real estate sectors.[30] While these examples are gathered from older cities in Europe, it is likely that with similar concerted strategies, high-speed rail could act as an anchor for development in newer, fast-growing regions like those in California and Texas, or play a virtually identical role in older cities in the Northeast and Midwest, where a high-speed rail station could reinvigorate the image of a city and anchor re-development strategies.

In summary, Ribalayguay and García found that high-speed rail's impact on urban revitalization is not automatic. Rather, only with a set of focused strategies, including setting aside land for new, intensified development and attracting such development with appropriate policies or enticements, has high-speed rail been shown to impact land-use change. We infer that the intensification of land uses that can result around high-speed rail stations would result in more-energy-efficient structures and travel patterns that also support walking, transit use, and reduced demand for auto trips. When viewed in this context, high-speed rail can be considered one additional strategy that complements city living and affirms city cores as the primary location for business activities, resulting in more-efficient transportation choices.

Conclusions and Potential for HSR to Contribute to Reductions in Oil Consumption and Emissions

High-speed rail has the potential to outperform other transportation modes in terms of oil consumption and emissions. In particular, it provides the advantage of electrification for intercity travel. However, any expectation regarding these benefits from high-speed rail deployment must be tempered by certain realities:

1. The overall potential for HSR to *directly* reduce transportation impacts is limited by the small fraction of travel taking place in appropriate intercity travel corridors compared to the overall market for passenger transportation.

2. In order to adequately reduce fossil-fuel consumption, trains must have high load factors and be sited in locations with high potential for ridership attracted from alternative modes. Such locations are likely to include dense, transit-connected corridors where there is a high propensity for passengers to switch to high-speed rail from other modes.

3. Electric-energy sources must be powered by renewable energy sources.

4. Criteria for assessing high-speed rail deployment should include the possibility of diminished advantage as technology evolves (particularly in the advent of improvements to and adoption of battery electric vehicles).

5. Compared to automobiles, train technology may have less opportunity for efficiency improvements. However, there is still much room for improved energy use per passenger-mile, particularly through the following:
   - Increased load factors
   - Regenerative braking

- Mass reduction
- Less frequent stops
- Lower maximum speed

Finally, the bulk of high-speed rail benefits in terms of energy and emissions may come from indirect effects on land-use, secondary trips, and so forth. These are areas for further research and should be seen as crucial to understanding the full public policy benefits of high-speed rail.

Notes

1. Governors Scott Walker of Wisconsin, John Kasich of Ohio, and Rick Scott of Florida turned back federal funding for rail projects in their states. The projects were to extend an existing conventional passenger rail service in Wisconsin, to upgrade a freight corridor in order to carry passenger trains in Ohio, and to build a new dedicated high-speed rail corridor connecting Tampa to Orlando in Florida.

2. Amtrak, "Amtrak Sets New Ridership Record" (news release, Oct 10, 2012), www.amtrak.com/ccurl/636/294/Amtrak-Sets-New-Ridership-Record-FY2012-ATK–12-092.pdf.

3. P. S. Hu, T. R. Reuscher, and US Department of Transportation, *Summary of Travel Trends—2001 National Household Travel Survey* (publication of the Federal Highway Administration, US Department of Transportation, Washington, DC, 2004), nhts.ornl.gov/2001/pub/stt.pdf; see also: Bureau of Transportation Statistics, *America on the Go—Findings from the National Household Travel Survey* (BTS publication, US Department Of Transportation, Washington, DC, 2006), www.bts.gov/publications/america_on_the_go/long_distance_transportation_patterns/.

According to the 2001 National Household Travel Survey, total annual passenger travel was approximately 4 trillion person-miles, of which approximately 1.3 billion person-miles (~33 percent) are categorized as long-distance, or greater than 50 miles. (Data Sources: NHTS 2001, *Summary of Travel Trends*, table 1; *America on the Go*, Findings from the NHTS 2001.)

4. Ray LaHood, "DOT Awards $2.4 Billion to Continue Developing 21st Century High-Speed Passenger Rail Corridors" (news release, US Department of Transportation, Washington, DC, 2010), fastlane.dot.gov/2010/10/dot-awards–24-billion-to-continue-developing–21st-century-high-speed-passenger-rail-corridors.html, accessed December 10, 2010; US Department of Transportation, *Vision for High-Speed Rail in America* (publication of the Federal Railroad Administration, US Department of Transportation, Washington, DC, 2009), www.fra.dot.gov/downloads/rrdev/hsrstrategicplan.pdf.

5. Most of this discussion relates to rolling vehicles—namely automobiles and railroad cars. Airplanes face fundamental physical limitations as well, but are governed by very different physical constraints.

6. A. G. Álvarez, "Energy Consumption and Emissions of High-Speed Trains," *Transportation Research Record: Journal of the Transportation Research Board* 2159, no. 1 (2010): 27–35, trb.metapress.com/index/V1R82Q772M218154.pdf.

7. R. U. Ayres, L. W. Ayres, and B. Warr, "Energy, Power and Work in the US Economy, 1900–1998," *Energy* 28, no. 3 (2003): 219–73, doi:10.1016/S0360-5442(02)00089-0.

8. R. Nolte, F. Würtenberger, and International Union of Railways, *Evaluation of Energy Efficiency Technologies for Rolling Stock and Train Operation of Railways* (publication of International Union of Railways, 2003), www.uic.org/download.php/environnement/energy_EVENT.pdf.

9. W. Gunselmann, "Technologies for Increased Energy Efficiency in Railway Systems," *2005 European Conference on Power Electronics and Applications* (report, Institute of Electrical and Electronics Engineers, 2005), 1–10, ieeexplore.ieee.org/xpls/abs_all.jsp?arnumber=1665902.

10. S. Hillmansen and C. Roberts, "Energy Storage Devices in Hybrid Railway Vehicles: A Kinematic Analysis," *Proceedings of the Institution of Mechanical Engineers, Part F: Journal of Rail and Rapid Transit* 221, no. 1 (2007): 135–43, doi:10.1243/09544097JRRT99; see also: H. Ishida and S. Iwakura, "Effects of High-Speed Transportation Systems on Environmental Improvement in Japan," *Japan Railway & Transport Review* 18 (December 1998): 12–20.

11. A. R. Miller and J. Peters, "Fuelcell Hybrid Locomotives: Applications and Benefits," *Proceedings of Joint Rail Conference 2006* (2006), 287–93.

12. M. Chester and A. Horvath, "Life-Cycle Assessment of High-Speed Rail: The Case of California," *Environmental Research Letters* 5, no. 1 (2010), doi:10.1088/1748-9326/5/1/014003.

13. E. Burgess, *Sustainability of Intercity Transportation Infrastructure: Assessing the Energy Consumption and Greenhouse-Gas Emissions of High-Speed Rail in the US* (master's thesis, Arizona State University, School of Sustainability, 2011).

14. J. J. Lee, S. P. Lukachko, I. A. Waitz, and A. Schafer, "Historical and Future Trends in Aircraft Performance, Cost, and Emissions," *Annual Review of Energy and the Environment* 26, no. 1 (2001): 167–200, doi:10.1146/annurev.energy.26.1.167.

15. Burgess, *Sustainability of Intercity Transportation Infrastructure*.

16. Ibid.

17. Ibid.

18. Ibid.

19. Since 1990, US passenger vehicle fleets have had scrappage rates on the order of 4–7 percent per year. Thus, fleet turnover time is likely to occur within 14–25 years. Consequently, a unique potential benefit of high-speed rail is that it could have a much more disruptive impact on intercity travel, since vehicle turnover time is not needed for passengers to switch to less energy-intensive rail modes.

20. For instance, a recent model of the Tesla Roadster is purported to achieve >300 miles on a single charge; see: www.teslamotors.com/models/facts.

21. For an example of battery-swap stations, see: Project BetterPlace, www.betterplace.com/the-solution-switch-stations.

22. J. Marriott, H. S. Matthews, and C. T. Hendrickson, "Impact of Power Generation Mix on Life-Cycle Assessment and Carbon Footprint Greenhouse-Gas Results," *Journal of Industrial Ecology* 14, no. 6 (2010), doi:10.1111/j.1530–9290.2010.00290.x.

23. P. Todorovich and Y. Hagler, *High-Speed Rail in America* (New York: America 2050, Regional Plan Association, 2011), www.america2050.org/2011/01/high-speed-rail-in-america .html.

24. Ibid.

25. Ibid.

26. P. Todorovich, D. Schned, and R. Lane, *High-Speed Rail: International Lessons for US Policy Makers*—Policy Focus Report (Cambridge, MA: Lincoln Institute of Land Policy, 2011).

27. L. Bertolini and T. Spit, *Cities on Rails* (London: E & F N Spoon, 1998).

28. C. B. Ribalayguay and J. G. García, "HSR Stations in Europe: New Opportunities for Urban Regeneration" (paper presented at the Fiftieth Anniversary European Congress of the Regional Science Association International, Jönköping, Sweden, 2010), www.scribd.com /doc/49254777/RIBALAYGUA-AND-GARCIA.

29. Ibid.

30. Ibid.

9 The Challenges and Benefits of Using Biodiesel in Freight Railways

SIMON MCDONNELL AND JIE (JANE) LIN

The recent history of biofuels, particularly biodiesel fuel, in the United States has been a turbulent one. After an initial boom in the production of biodiesel and a huge expansion in capacity in advance of expected increases in demand, the bottom fell out of the market by the end of 2010. Annual biodiesel production, which had approached 700 million gallons in 2008, fell to just over 500 million gallons by 2009 and reached the lowest point at only 340 million gallons in 2010.[1] Since then, however, the boom times have returned; 2011 saw the highest-ever level of production, at almost a billion gallons.[2] Much of this ebb and flow has been caused by the policy environment faced by biodiesel producers. Since the 1970s, federal efforts have broadly favoured increased production of biofuels, particularly corn-based ethanol. Since the 1990s, those encouragement efforts have expanded to include biodiesel production.[3] The enactment of the Energy Policy Act of 2005 and, more recently, the Energy Independence and Security Act of 2007 have mandated increased biofuel penetration into the transportation fuel market.[4] By diversifying the transportation fuel mix, federal policies aim to improve energy security and environmental performance, although the latter is more controversial than the former.[5] Together, these actions have created a market for biofuels as a substitute for its petroleum counterparts. Of particular interest is the market for biodiesel, an overshadowed alternative to the more popular corn-based ethanol gasoline substitute.

Indeed, biodiesel currently only accounts for approximately 10 percent of the US biofuels market, which is mostly made up of ethanol fuel. While much of the policy focus has been on developing a market for biofuels, primarily corn-based ethanol, for transportation engines there is a potential for biodiesels to play a significant part in reducing consumption of petroleum diesel, particularly in the non-road engine sector. For instance, the US rail industry currently consumes an amount approaching 4 billion gallons of petroleum diesel per annum. This translates to approximately an 8.5 percent share of total US diesel transportation consumption.[6] It can be hypothesized that a large penetration of biodiesel into this market will have significant impacts both on rail energy consumption and on sectors supplying biodiesel inputs. This is particularly relevant given the negligible contribution that biodiesel currently makes to the fuel mix for rail freight. Furthermore, recent research and practical experience have indicated that biodiesel, especially when blended in low proportions with petroleum diesel (i.e., 20 percent or less), can be technologically compatible with most existing locomotive fleets without losing much of the power performance—a 1–2 percent reduction depending on the level of biodiesel fuel use.[7]

Indeed, it can be argued that given the low levels of fleet turnover—locomotives are often in operation for 50 years or more—fuel-replacement strategies offer the best chance of changing the energy-consumption mix of the sector and that biodiesel substitution may act as that driving force. Therefore, any low-cost strategy that replaces petroleum fuel with a renewable substitute in existing locomotives may very well be attractive to both rail operators and policy makers aiming to reduce reliance on crude oil extracts. In fact, although efforts are still in the experimental phase, the US railroad sector has begun to explore biodiesel penetration options; for instance, the Society of Automotive Engineers are investigating the technical feasibility of incorporating different blends of biodiesel into the rail fuel mix, and some freight rail operators have piloted biodiesel experiments in recent years.[8]

The important question here is whether biodiesel (especially when blended with petroleum at low to medium levels) is truly a "low-hanging fruit" solution to reducing fossil-fuel energy use and whether it can mitigate other externalities associated with current practices (e.g., emissions of local pollutants and greenhouse gases). Ultimately, are policy incentives justified in making biodiesel fuel more economically competitive with petroleum-based fuels?

The question is particularly relevant because the cost of producing biodiesel remains an issue. The recent trends for the cost of producing both biodiesel and its petroleum counterpart have been unsettled but generally upward. For instance, rail fuel costs, consisting mostly of petroleum diesel, quadrupled between 2004 and late 2008 before collapsing in 2009. But by the end of 2012, these costs were more than three times their 2004 levels.[9] Meanwhile, the soybean-oil trading prices—soybeans are the primary feedstock for biodiesel in the United States—grew by more than 150 percent between 2004 and 2008 before suffering a similar collapse. By April 2012, soybean prices had almost returned to their 2008 peak, trading at near-historical highs. In general biodiesel remains more costly than its petroleum counterpart. The production cost of biodiesel has traditionally been as much as three times that of gasoline and petroleum diesel.[10] It is likely that if biodiesels are to form a major part of the rail fuel mix, significant policy incentives will be required.[11]

The environmental performance of biodiesel fuel in terms of criteria pollutants and greenhouse gases is another issue.[12] Research has shown that biodiesel used in 50 percent and 100 percent blends (referred to as B50 and B100, respectively) does have the potential to reduce emissions of carbon monoxide (CO) and hydrocarbons (HC), but this comes at the cost of increased nitrogen oxides (NO_x), a precursor to ground-level ozone (so-called bad ozone) formation.[13] In particular, higher blends of biodiesel (e.g., 100 percent) are estimated to increase NO_x emissions by 2–5 percent. Lower blends will have more modest impacts on NO_x emissions. This undesirable outcome is inconsistent with the new US EPA emission standards for locomotives manufactured after January 1, 2000.[14] On the other hand, carbon dioxide (CO_2) shows no difference in tailpipe exhaust between petroleum diesel and biodiesel.[15] In fact, research has shown that from fuel life-cycle perspective—that is, from growing and harvesting the feedstock to production and distribution of biodiesel fuel, to final consumption of the fuel—biodiesel fuel actually provides CO_2 benefits resulting from the renewability of the biodiesel, because the crops take in CO_2 during photosynthesis.[16]

However, such analyses do not take into account the land-use change due to farmers' response to higher crop demand and prices by converting non-arable land to new cropland, which may result in increase of carbon emissions.[17] Moreover, if there was a large penetration into the rail freight market, what impact would it have on land use? If, for instance, as is the case now, most US biodiesel comes from soybeans, will a large-scale penetration have a significant impact on land devoted to growing soybeans? Will this have an impact on the price of soybeans and, more broadly, food prices as fuel production potentially squeezes out other uses? How would this affect biodiversity on those impacted lands? All these questions point to a larger question: How reliable is biodiesel as a fuel source?

To begin addressing these questions, this chapter looks at the potential energy use and environmental impacts of a large-scale biodiesel penetration into the US rail industry and the wider impacts on land use and food supply. The investigation and discussion are focused on the national scale and are restricted to Class I freight rail, which accounts for 80 percent of non-car shipping rail cars of national rail operations in the United States.[18] Admittedly, this chapter will only scratch the surface of the biodiesel controversies. There remains a paucity of research related to the role of biodiesels in reducing energy usage in the rail industry specifically and the subsequent wider impacts implied by such a penetration of soybean-based biodiesel in a large transportation sector. We are in great need of more research efforts.

Potential Impact of Biodiesel Fuel on Energy Consumption in Freight Rail

In 2000, as little as 2 million gallons of biodiesel were produced in the United States. By 2006, this figure had risen to 250 million gallons, and after recent fluctuations, 2011 saw a record year for biodiesel production with almost a billion gallons produced.[19] As recently as 2008, the Energy Information Administration of US Department of Energy (DOE) forecasted total biodiesel use in the United States in 2030 at 1.3 billion gallons—accounting for 1.6 percent of total diesel consumption by volume, a figure that seems to be within reach much sooner if recent trends continue. In addition to biodiesels from more traditional sources (i.e., soybeans), advanced biodiesels (i.e., diesel liquids produced from biomass—BTL) are expected to account for 4.2 billion gallons by the same time—just under 5 percent of total diesel consumption.[20]

Before understanding the impacts of any potential biodiesel penetration in Class I freight rail, let us first look at the current status of the Class I freight rail (table 9.1).[21] Over the period of 1995–2005, the number of train-miles grew by 1.8 percent per annum while freight car-miles grew by 2.2 percent per annum. This indicates an increase in the length of trains (i.e., number of cars) over the period. A revenue ton-mile is one ton of freight carried over one mile by train and is the key indicator used by railways in determining the efficiency of the mode. Between 1995 and 2005, the industry experienced an almost 3 percent per annum growth in revenue ton-miles. At the same time, the energy intensity—the energy needed to carry a ton-mile (measured as Btu/ton-mile)—has fallen by 1 percent per annum. Total energy use (measured in million Btu) has grown by 1.6 percent over the same period. Extrapolating that historical trend of 1.6 percent growth, it means that energy demand for Class I freight rail would grow from 546 trillion Btu in 2005 to 691.56 trillion Btu in 2020, which correspond to 4.2 and 5.3 billion gallons, respectively, of the no. 2 petroleum diesel that US locomotives typically use.[22] Even using a more conservative growth rate of 1 percent according to the DOE's Energy Information Administration projection, by 2020 energy demand would grow to 652.89 trillion Btu, representing a demand for 5.0 billion gallons of no. 2 petroleum diesel.

Potential Energy Consumption Benefits of Biodiesel in Freight Rail

For the purposes of this chapter, we consider the potential energy-consumption benefits of biodiesel from the entire fuel life-cycle, which provides a holistic and complete

Table 9.1 Summary statistics for Class I freight railroads, 1970–2005[a]

Year	Train-miles[b] (millions)	Car-miles[b] (millions)	Revenue ton-miles (millions)	Energy intensity (Btu/ton-mile)	Energy use (trillion Btu)
1970	427	29,890	764,809	691	528.1
1975	403	27,656	754,252	687	518.3
1980	428	29,277	918,958	597	548.7
1985	347	24,920	876,984	497	436.1
1990	380	26,159	1,033,969	420	434.7
1995	458	30,383	1,305,688	372	485.9
2000	504	34,590	1,465,960	352	516.0
2005	548	37,712	1,696,425	337	571.4
Average annual percentage change					
1970–2005	0.7%	0.7%	2.3%	−2.0%	0.2%
1995–2005	1.8%	2.2%	2.7%	−1.0%	1.6%

[a] Adapted from Association of American Railroads, *Railroad Facts*, 2005 Edition (AAR publication, Washington, DC, November 2006), 27, 28, 33, 34, 36, 49, 51, and 61; cited in Oak Ridge National Laboratory *Transport Energy Data Book*, 26th ed. (ORNL publication, 2007).

[b] Train-miles and car-miles measure the distance traveled by trains and all the cars. For instance, a train with 40 cars traveling 1 mile is measured as one train-mile and 40 car-miles.

picture of total energy consumption. Biodiesel can be manufactured from a variety of renewable biomass feedstocks. As noted, currently most biodiesel is produced from soybeans; the feedstock accounts for 70–95 percent of all biodiesel produced in the country.[23] As such, this chapter considers only soybean-based biodiesel fuel. There are different blending options of biodiesel in petroleum diesel and these thus form different biodiesel fuels whose chemical properties are different. The common volume blending includes 20 percent (B20) and 80 percent (B80) of biodiesel, respectively, blended with petroleum diesel. B20 is viewed as the lowest biodiesel blending that can deliver energy benefits with minimum adverse environmental effects over petroleum diesel without compromising engine performance. It is reasonable to assume that no technological alteration takes place and that the cost factor to switch is eliminated. In other words, B20 may be the "lowest-hanging fruit" in the biodiesel fuel family and thus the strategy with perhaps the greatest chance of occurring. We estimated the annual total life-cycle energy demand[24] associated with petroleum diesel and B20 use in rail nationally between 2005 and 2020.[25] We assumed a market that is all soybean-based with production that uses existing technology. This assumption will likely ignore the potential of market penetration of the "second generation" biodiesel (also called cellulosic biodiesel or biomass-to-liquid diesel). However, given the large uncertainty regarding the marketability of these fuels and lack of empirical data, we focus our discussion only on existing technology.

Assuming low (20 percent), moderate (50 percent), and high (100 percent) levels of B20 biodiesel penetration into the national Class I freight rail market with a 1 percent annual growth rate of energy demand between 2010 and 2020, even a low penetration of B20 requires over 200 million gallons of biodiesel by 2020. With a 1.6 percent annual growth rate of energy demand, the requirement rises to 212 million gallons by 2020. With a 50 percent penetration of B20 under the 1 percent growth assumption, an additional 301 million gallons per annum will be demanded by 2020. With a 1.6 percent growth rate, the additional requirement rises to 319 million gallons. If we hypothesize 100 percent B20 penetration in the sector by 2020, biodiesel demand exceeds 1 billion gallons by 2020. With a 1.6 percent annual growth rate, this demand rises to 1.06 billion gallons. In other words, the biodiesel share of the total Class I freight rail diesel fuel demand in 2020 may range between 4 percent and 20 percent.

The resulting total life-cycle energy demand in the complete (100 percent) B20 penetration scenario, which represents the maximum possible energy demand, increases slightly, by 2.5 percent in 2015 and 4.9 percent in 2020, over that of the no-biodiesel-penetration case, as shown in table 9.2.[26] More important, the fossil-fuel energy requirements are reduced by 11–13 percent—roughly the amount of energy consumption by the entire passenger rail of the nation in 2004–2005.[27] This obviously represents a desirable scenario from the perspective of reducing dependence on fossil fuel. However, this moderate reduction would need to be achieved by 100 percent B20 penetration in a ten-year time frame, which will obviously require a favorable policy environment and very strong incentives, something that is largely absent from the policy arena right now. It is reasonable to expect a much smaller reduction to be achieved by 2020 with a much lower percent penetration of B20.

Biofuel Energy Policy Incentives in the United States

In recent years, energy policy has come to be focused on the overarching aims of increased supply security and climate-change mitigation. In the United States, the first federal intervention aimed at encouraging the development of a market for biofuels was the Energy Security Act of 1979, which created tax credits for firms selling or using alcohol as a fuel.[28] Later, the passing of the Clean Air Act Amendment of 1990 (CAAA) and more specifically the Energy Policy Act of 1992 (EPACT 1992) incentivized the recent growth in biofuel use and production. The rationale for the latter was centered on concerns over supply security and it first mandated alternative fuel use in certain federal fleets; however, biodiesels were specifically excluded. It was only

Table 9.2 Life-cycle energy demands in no-biodiesel and complete-B20-penetration scenarios (both at 1 percent energy-demand growth rate)

	0% biodiesel (trillion Btu)		100% B20 penetration (trillion Btu)	
	Total energy	Fossil fuel energy	Total energy	Fossil fuel energy
2005	659.76	657.66	659.76	657.66
2010	703.65	701.55	703.65	701.55
2015	751.45	749.32	769.97	651.07
2020	790.22	788.10	829.02	703.73

with the 1998 reauthorization of the act that the incentives were extended to classes of biodiesel—a possible explanation for the relative dominance of ethanol in the US biofuel mix, along with the higher yields from corn than from soybeans per acre.

Production of biodiesel has also been incentivized through a number of schemes such as the American Jobs Creation Act (1994), which introduced a biodiesel tax credit aimed at lowering the production costs of blenders mixing biodiesel with petroleum diesel. The credit amounted to a penny percentage point for vegetable oil, that is, blenders with a B20 mix received a 20-cent-per-gallon excise tax credit. The Energy Policy Act of 2005 extended this incentive to the end of 2008 by the inclusion of the Renewable Fuels Standard, which mandates the minimum renewable fuel share to be included in the national gasoline usage pool—an amount that increases each year. The provisions include a credit-trading scheme among blenders. Further supply-side incentives included cash support from the US Department of Agriculture Commodity Credit Corporation bioenergy program to reimburse farmers for growing plants that produce biofuels. This support was phased out at the end of financial year 2006.

The succeeding Energy Independence and Security Action of 2007 heightens the policy focus on biodiesel and outlines ambitious longer-term targets for biofuel and biodiesel penetration, including the increase of the Renewable Fuels Standard to 36 billion gallons by 2022.[29] Of this, 21 billion will be made up of "advanced biofuels" through improvements in technology, and 16 billion will come from cellulosic (second-generation) biofuels. The Renewable Fuels Standard requires biodiesel and advanced biofuels deliver a life-cycle 50 percent reduction of GHGs compared to that of their petroleum counterpart.

Impact on Land Use

More recently, land-use impacts of increased biofuel production and demand have gained increasing attention from both policy and academic perspectives.[30] In the

United States, ethanol has already had a tremendous impact on the number of acres under corn—in 2007, there was a 15 percent increase to 90 million acres.[31] Some projections envisage corn ethanol in 2016 using 43 percent of the US corn land for grain.[32] Acres under soybeans fell from 75 million in 2006 to 63.6 million in 2007.[33] As demand has recovered for biodiesel, recent trends have seen a rally in the number of acres under soybeans; by 2011, the numbers had recovered to 75 million acres.[34]

Overall, land-use impacts arising from soybean production associated with increased biodiesel production are generally not quantified,[35] and the potential effect of biodiesels on land-use change is an area demanding more attention and research. The limited studies have shown that biomass-based fuel penetration in light-duty vehicles has low energy efficiency and high land demand compared to renewable electricity-based fuel cycles (with hydrogen and battery electric vehicles).[36]

Using our previous investigation of energy demand for Class I freight rail biodiesel penetration, we find that a 20 percent B20 penetration by 2020, under the assumption of 1 percent growth of energy demand, would require 2.8 million acres of additional land per annum—relatively modest in the context of approximately 75 million acres presently under soybeans per annum.[37] With a 50 percent B20 penetration, soybean land requirement increases to approximately 7 million per annum, accounting for almost 10 percent of land currently under soybeans. As expected, a complete switch to B20 fuel by 2020 presents a considerable impact on land use, as the total additional acreage required to support this switch increases to approximately 14.2 million acres under the assumption of 1 percent growth rate of energy demand, or 15 million acres with a 1.6 percent growth rate. So it is likely that, without significant technological improvements (e.g., through viable second-generation biodiesel being brought to market), such a demand by the rail freight industry would impact significantly on the biodiesel and soybean industry.

Increased demand for land due to biofuels may incentivize conversion of land currently used for alternative purposes (e.g., forestry or arable land). While not directly impacting food prices (see below), the likely negative environmental impact on, for instance, biodiversity and carbon release from land may be significant. It was concerns such as these that led to the European Union scaling back on goals for biofuel penetration in the European transportation sector.[38]

Finally, there may be potential to grow biofuel crops on degraded or otherwise unsuitable land. The International Union of Food Science and Technology (IUFOST) notes that there are certain biofuel crops that are relatively drought-resistant and are able to grow on degraded land. However, as of yet, there is little research on yields and the impact on soil.

Impact on Food Supply

Biodiesel production holds significant advantages over ethanol due to lower agricultural impacts and greater efficiency in fuel conversion associated with the former.[39] However, even modest replacement of petroleum will likely impact food supplies in the United States and beyond as producers respond to different policy and market incentives.

Despite this, it is worth noting that food demand for soybean oils had been stagnating as a result of increased concerns about trans-fatty acids.[40] Perhaps the increase in biodiesel demand may have come at a good time for soybean producers. And, as recently as 2008, demand for biodiesel accounted for only 15 percent of total soybean oil use.[41] In addition, much of soybean production is for non-domestic consumption; about 45 percent of all US-grown soybeans are exported. This percentage has remained constant over the last five years even with the fluctuations in the US market.[42] After the initial squeeze on acreage as a result of the corn ethanol boom, soybean acreage has been robust to corn competition. That and the high proportion of exported soybeans suggest that there remains some flexibility in the US production market. As a result, if demand continues to ramp up, the impact may be felt more on the export market than on domestic food prices, initially at least. Despite that, and as already noted, the direct impact of the biodiesel boom on soybean prices has been significantly upward. The real question is whether the increase in soybean demand currently projected (and the one hypothesized in our analysis) indirectly drive up other food prices as other land uses get squeezed.

Food demand is also a moving target, due to population increases and expected economic development in poorer parts of the world. Demand for food, like water and energy, is expected to rise by up to 50 percent by 2030. Because of these factors, any restriction on the supply of food crops for human consumption, in the United States or abroad, is likely to impact prices. A recent World Bank report estimated that 70 percent of food-price increases were the result of increased biofuel demand and that over three in every ten pounds of corn grown in the United States go to producing ethanol.[43] Less is known about biodiesel, but in the European Union over half of all vegetable oils already go to biodiesel production.[44] How food prices are impacted will likely be determined by how food producers respond. They can continue to utilize the supply of existing agricultural land, or they can, where possible, expand the supply of agricultural land through land-use change, possibly using marginal land that would not be otherwise suitable for crop production.[45]

Regarding the first point, the issue comes down to whether growing biofuel crops competes with or complements existing patterns. Evidence from Brazil indicates that high energy-price fluctuations eventually translate into higher food-price variation. On the other hand, some have argued that growing biofuel crops in tandem with food cash crops will actually increase resource-use efficiency.[46] For instance, soybean yields have been growing as technology and practices improve. The US Department of Agriculture estimated that each soybean acre yielded an average of 41.5 bushels in 2007, a rate that has been growing at 0.45 bushels per annum. Yields are expected to be 46.6 bushels per acre by 2016.[47]

Impact on GHG Emissions and Other Pollutants

Again, with 100 percent B20 penetration in Class I freight rail, the reduction in CO_2 equivalent emissions (including three major GHGs—CO_2, CH_4, and N_2O) is notable (see table 9.3), particularly reductions associated with the so-called well-to-pump stage—processes and activities during fuel production (including feedstock production) and distribution.[48] The well-to-pump CO_2 equivalent reduction could be as high as 60 percent if the penetration of B20 reaches 100 percent by 2020. However, the overall life-cycle reduction would be a modest 9.8 percent. There would be a GHG benefit but not by a large margin. On the other hand, there could be a considerable increase in volatile organic chemicals (VOC), roughly 40 percent over life cycle, or 3.7 times during well-to-pump if there were 100 percent B20 penetration by 2020. Considering that VOC is a precursor to ground-level ozone formation, this large increase in VOC has an important implication on local ozone pollution and subsequently on public health. For the other pollutants, there is no benefit gained, however, nor is there obvious worsening.

Conclusions and Policy Implications

The impact of biodiesel penetration into the Class I freight rail sector is particularly relevant in the present environment as a result of the increased policy focus on reducing energy dependence and emissions from non-road mobile sources and, more recently, the concern about the potential land-use impacts of biofuels. Because of the long-life nature of locomotive engines, any regulatory impacts related to their vehicle cycle would have only marginal effects in the medium term. Train operators will likely be unwilling to engage in fleet-replacement strategies without very strong incentives. On the other hand, we hypothesize that fuel-based interventions may potentially be a "low-hanging fruit." By presenting rail freight operators with low- or no-cost alternatives to existing practices, there could be scope for a relatively large change in behavior consistent with the federal government's larger environmental and energy security goals. But, because of recent increases in soybean prices and the continued relative

Table 9.3 Emissions for no-biodiesel and complete-B20-penetration scenarios (both at 1 percent energy-demand growth rate)[a]

| | 0% biodiesel | | | | 100% B20 penetration | | | |
| | 2015 | | 2020 | | 2015 | | 2020 | |
	WTP[b]	LC[b]	WTP	LC	WTP	LC	WTP	LC
CO_2 eq. (mill. tons)	11.732	71.229	12.304	74.884	8.238	67.734	4.945	67.525
VOC (ktons)	4.836	45.421	5.057	44.732	13.826	54.411	23.873	63.548
CO (ktons)	8.210	139.038	8.515	146.124	9.327	140.156	10.171	147.780
NO_x (ktons)	25.844	746.834	25.950	730.569	28.025	749.015	29.425	734.044
PM_{10} (ktons)	5.869	31.175	6.027	30.636	6.141	31.447	6.498	31.107

[a] For both scenarios, emission results for years 2005 and 2010 are the same and skipped for comparison.
[b] WTP = well-to-pump, LC = life cycle

lack of economic competitiveness of biodiesel as an alternative fuel, the question will come down to federal willingness to intervene in the market.

However, we find only moderate evidence for the low-hanging-fruit hypothesis. Even the most optimistic penetration scenario results in a fall of 10–13 percent in life-cycle fossil-fuel energy demand. While this is equivalent to approximately the amount of energy consumed by the entire US passenger rail sector for 2004–2005, it represents only a modest decline in the grander scheme. What is more, the total life-cycle energy demand actually increases slightly. The resultant positive impact on GHGs is also modest—by 2020 emissions are 90.4 percent of what they would have been if no biodiesel penetration were to take place. However, these declines in energy consumption and the reduction of CO_2 equivalents would come as a result of complete penetration of B20 by 2020. In reality, the biodiesel penetration is likely to be more modest, and its mitigation effect in GHGs is likely to be even more marginal. As a result, it is difficult to conclude that a modest penetration of B20 diesel (likely supported by financial incentives) will be an especially effective tool in reducing fossil-fuel consumption and GHG emissions.

The impacts of such a penetration would also be felt beyond rail freight operators and soybean producers. For instance, such a large-scale penetration will likely have a significant impact on land devoted to growing soybeans in the United States With a 20 percent penetration rate, annual additional land requirements increase by almost 3 million acres—roughly 4 percent of present soybean land demand. However, a 100 percent penetration rate increases the additional land requirement to about 14 million acres—approximately a fifth of soybean acreage. As a result, the complete-penetration scenario will likely see significant pressures on existing land uses and soybean demand—possibly resulting in even more upward price pressures, possibly both at home and abroad. Food prices may increase as other crops are squeezed out by the soybean boom. Biodiversity and carbon estimates will also be negatively impacted if soybean producers respond to increased demand by converting other land uses to land for soybeans.

The complete-penetration scenario also implies almost a billion gallons of biodiesel demand from this sector by 2020, accounting for the majority of projected biodiesel production. This will potentially undermine the economics of B20 substitution, especially when the impacts on local pollutants and CO_2 emissions are marginal.

That is not all. CO_2 reductions will likely come at the price of higher criteria-pollutant emissions. In fact, with 100 percent penetration, we see VOCs register a 40 percent increase over the base-case scenario by 2020. Given that VOC is a precursor to ground-level ozone formation, this is a particularly worrying finding. While the remaining three criteria pollutants experience only moderate rises (0.5–1.5 percent), the trend is not positive. As the EPA policy agenda is presently focused on reducing emissions of PM10 and NO_x from non-road mobile sources, the finding that this policy adds nothing to achieving these objectives and may, in fact, result in even higher emissions will likely act as a further barrier to its adoption. In addition, the harmful local impacts of these pollutants would be expected to be more of an immediate concern to policy makers and rail operators, especially where operations are concentrated, such as in urban areas.

All these considerations point negatively to the question of how reliable biodiesel will be as a fuel source in the face of ramped-up demand. On a brighter note, the potential impact of second-generation biodiesels, with a large-scale market penetration, could greatly increase the efficiency of biodiesel blends and reduce the resultant land impacts. However, given the high degree of uncertainty about these fuels (both in terms of marketability and efficiency improvements), projecting the potential impacts remains difficult.

Indeed, we have alluded to only the cost barriers hindering B20 blend penetration and hypothesized a regulatory framework aimed at allowing it to compete successfully. However, strategies concerned with reducing the fossil-fuel energy consumption and the environmental impact of rail freight transportation must first be cognizant of whether such interventions are justified in the first place. This is particularly relevant given the increasing costs of biodiesel. Of course, such problems are not unique to the freight rail sector; however, the long-life characteristics of rail locomotives reduce the potential impact of alternative vehicle-cycle technological advances in the short to medium term. This again highlights the problems facing policy makers regarding this sector; as noted, EPA emissions standards apply to only new and remanufactured locomotives, while existing locomotives are largely exempt from such emissions reductions. Our findings show that the use of soybean-based biodiesel, aided by significant interventions in the policy sphere, will offer only moderate opportunities for emissions mitigation, in tandem with potentially significant downsides. Such concerns will only heighten the role of the EPA in reducing emissions from existing locomotives operations. In the absence of such intervention, economically competitive second-generation biodiesels will likely be an important tool for policy makers in the coming decade.

Notes

1. M. Carriquiry, "US Biodiesel Production: Recent Developments and Prospects," *Iowa Agriculture Review* 13, no. 2 (2007): 8–11.

2. US Energy Information Administration, "Monthly Biodeisel Production Report" (EIA report, US Department of Energy, Washington, DC, May 4, 2012), www.eia.gov/biofuels /biodiesel/production/.

3. National Research Council, *Renewable Fuel Standard: Potential Economic and Environmental Effects of US Biofuel Policy* (Washington, DC: The National Academies Press, 2011).

4. The Renewable Fuel Standard was created under the Energy Policy Act of 2005 and expanded under the Energy Independence and Security Act of 2007. It mandates a minimum volume of renewable fuel in the US transportation fuel mix; see: www.epa.gov/otaq/fuels /renewablefuels/index.htm.

5. United Nations Educational, Scientific and Cultural Organization/Scientific Committee on Problems of the Environment / United Nations Environment Programme, *Biofuels and Environmental Impacts: Scientific Analysis and Implications for Sustainability* (UNESCO, SCOPE, UNEP Policy Brief No. 9, June 2009), unesdoc.unesco.org/images/0018/001831/183113e.pdf.

6. S. C. Davis and S. W. Diegel, eds., "Energy Efficiency and Renewable Energy," *Transportation Energy Data Book*, 25th ed. (publication of the Oak Ridge National Laboratory, prepared for the Office of Planning, Budget Formulation and Analysis, US Department of Energy, Washington, DC, 2006); see also: Bureau of Transportation Statistics *National Transport Statistics 2011* (BTS publication, Research and Innovative Technology Administration, US Department of Transportation, Washington, DC, 2011), www.bts.gov/publications/national_transportation_statistics/pdf/entire.pdf.

7. S. G. Fritz, "Evaluation of Biodiesel Fuel in an EMD GP38–2 Locomotive" (report no. NREL/SR–510–33436, Subcontractor Report for the National Renewable Energy Laboratory, Southwest Research Institute, San Antonio, Texas, May 2004); see also: F. Stodolsky, *Railroad and Locomotive Technology Roadmap* (publication no. ANL/ESD/02–6, Argonne National Laboratory, Transportation Technology R&D Center, 2002); R. Dunn, "Biodiesel as a Locomotive

Fuel in Canada" (report prepared for Transportation Development Centre of Transport Canada, TP14106E, May 2003); International Union of Railways, *Railways and Biofuel* (UIC report—final draft, produced in cooperation with the Association of Train Operating Companies, July 2007).

8. National Biodiesel Board, "NBB Technical Team Collaborates with the Railway Industry," *Biodiesel Magazine*, October 25, 2011, biodieselmagazine.com/articles/8134/nbb-technical-team-collaborates-with-railway-industry; see also: Dunn, "Biodiesel as a Locomotive Fuel in Canada."

9. Association of American Railroads, *Monthly Railroad Fuel Price Indexes* (AAR publication, Washington, DC, June 2008), www.aar.org/PubCommon/Documents/AboutTheIndustry/Index_MonthlyFuelPrices.pdf; Association of American Railroads, *Monthly Railroad Fuel Price Indexes* (AAR publication, Washington, DC, April 2012), www.aar.org/~/media/aar/RailCostIndexes/MRF201204.ashx; Research and Innovative Technology Administration, *Index of Railroad Fuel Prices* (publication of the Bureau of Transportation Statistics, US Department of Transportation, Washington DC, April 2012), www.bts.gov/publications/multimodal_transportation_indicators/april_2012/html/rail_fuel_price.html.

10. L. Fulton, "Biofuel Costs and Market Impacts in the Transport Sector," *Energy Prices & Taxes* (publication of the International Energy Agency, 1st Quarter 2005), xi–xxvi.

11. Index Mundi, "Soybeans Daily Price (10-year trend)/Soybeans Monthly Price—US Dollars per Metric Ton" (report updated daily), www.indexmundi.com/commodities/?commodity=soybeans&months=120.

12. *Criteria pollutant* refers to any of six pollutants regulated by the US Environmental Protection Agency (US EPA), i.e., carbon monoxide (CO), nitrogen oxides (NO_x), lead (Pb), ozone (O_3), particulate matter with diameter less than 10 micrometer (PM10) and less than 2.5 micrometer (PM2.5), and sulfur dioxide (SO_2).

13. C. Grimaldi, L. Postrioti, M. Battistoni, and F. Millo, "Common Rail HSDI Diesel Engine Combustion and Emissions with Fossil/Bio-Derived Fuel Blends" (SAE Paper no. 2002-01-0865, *Proceedings from the SAE 2002 World Congress*, Detroit, MI, USA, March 4–7, 2002).

14. As promulgated by the EPA in 1998; see: www.epa.gov/otaq/regs/nonroad/locomotv/frm/f99037.pdf.

15. US Environmental Protection Agency, *A Comprehensive Analysis of Biodiesel Impacts on Exhaust Emissions* (draft technical report no. EPA420-P-02-001, EPA, Washington, DC, October 2002); see also: J. Sheehan, V. Camobreco, J. Duffield, M. Graboski, and H. Shapouri, *Life-Cycle Inventory of Biodiesel and Petroleum Diesel for Use in the Urban Bus* (final report prepared for US Department of Energy, Office of Fuels Development, and US Department of Agriculture, Office of Energy; National Renewable Energy Laboratory, NREL/SR–580–24089, May 1998).

16. EPA, www.epa.gov/otaq/regs/nonroad/locomotv/frm/f99037.pdf.

17. Ibid.; see also: T. Searchinger, R. Heimlich, R. A. Houghton, F. Dong, A. Elobeid, J. Fabiosa, S. Tokgoz, D. Hayes, and T. Yu, "Use of U.S. Croplands for Biofuels Increases Greenhouse Gases through Emissions from Land-Use Change," *Science* 319 (2008): 1238–40.

18. Association of American Railroads, *Railroad Facts, 2005 Edition* (AAR publication, Washington, DC, November 2006), 27, 28, 33, 34, 36, 49, 51, 61.

19. EIA, "Monthly Biodeisel Production Report."

20. US Energy Information Administration, *Annual Energy Outlook 2008 Overview: Energy Trends to 2030* (EIA publication, US Department of Energy, Washington, DC, 2008), www.eia.doe.gov/oiaf/aeo/pdf/overview.pdf.

21. The Interstate Commerce Commission designates Class I railroads on the basis of annual gross revenues. In 2003, there were seven railroads with this designation.

22. The energy content of diesel fuel varies slightly by type of diesel fuel. The difference between the energy content of biodiesel and petroleum diesel can be significant, especially at levels approaching B100 (US Department of Energy, 2006). The latter is between 5 and 8 percent less energy dense, and fuel efficiency is about 2 percent less than diesel (Fritz, 2004). 130,000 btu/gallon of diesel is the average value adopted by US Department of Energy; see: www.eere.energy.gov/afdc/pdfs/afv_info.pdf; see also: US Department of Energy, *Biodiesel Handling and Use Guidelines*, 3rd ed. (DOE publication no. DOE/GO–1–2–6–2358, Washington, DC, September 2006).

23. Carriquiry, "US Biodiesel Production."

24. The life cycle can be disaggregated into the "well-to-pump" and the "pump-to-wheels" stages. In the former, we adopt many of the assumptions in the Argonne National Laboratory's GREET model (Greenhouse Gases, Regulated Emissions, and Energy Use in Transportation) version 1.8. These assumptions include petroleum-diesel pathway options, soybeans-to-biodiesel pathway options, the North American electricity grid mix, and associated energy efficiencies in farming and harvesting (for soybean feedstocks), extraction, refinery, production, and distribution (Huo et al., 2008). We also adopt the well-to-pump GHG and criteria-pollutant-emission rates in GREET. In the latter, we project growth rates in the pump-to-wheels energy consumption by Class I rail freight to calculate the absolute demand for petroleum diesel and biodiesel under our assumptions. See McDonnell and Lin, "The Use of Biodiesel in Railways," for technical details (below).

25. S. McDonnell and J. Lin, "The Use of Biodiesel in Railways and Its Impact on Greenhouse-Gas Emissions and Land Use" (TRB paper no. 09–1537, *Proceedings [Compendium DVD] of the 88th Transportation Research Board Annual Meeting*, Washington, DC, January 11–15, 2009).

26. It is assumed that biodiesel use in Class I rail freight will grow from a base level of 0 percent in 2010 to 100 percent in 2020 at an annual growth rate of 10 percent per annum, and that soybean yield per acre is 41.5 bushels and gallons of biodiesel from one acre of soybeans is 62.11 gallons.

27. S. C. Davis and S. W. Diegel, eds., *Transportation Energy Data Book*, 26th ed. (publication of the Oak Ridge National Laboratory, prepared for the Office of Planning, Budget Formulation and Analysis, US Department of Energy, Washington, DC, 2007).

28. F. Beck and E. Martinot, "Renewable Energy Policies and Barriers," *Encyclopedia of Energy*, ed. Cutler J. Cleveland (Oxford: Academic Press/Elsevier Science, 2004).

29. EIA has recently noted that the target will be missed due to uncertainty over technological developments and market penetration prior to 2022. A modified target of 32.5 billion gallons has been proposed. (Source: EIA Administrator Guy Caruso testimony to US Senate Committee on Energy and Natural Resources, March 4, 2008; see: www.planetark.org/daily newsstory.cfm/newsid/47352/newsDate/5-Mar–2008/story.htm.)

30. EPA, *Comprehensive Analysis of Biodiesel Impacts*; see also: D. C. Morton, R. S. DeFries, Y. E. Shimabukuro, L. O. Anderson, E. Arai, F. del Bon Espirito-Santo, R. Freitas, and J. Morisette, "Cropland Expansion Changes Deforestation Dynamics in the Southern Brazilian Amazon," *Proceedings of the National Academy of Sciences of the United States of America (PNAS)* 103, no. 39 (2006): 14637–41.

31. US Department of Agriculture, *Agricultural Projections to 2016* (publication of the US Department of Agriculture, Economic Research Service, Washington, DC, February 2007).

32. Fulton, *Biofuel Costs*.

33. US Department of Agriculture, *Agricultural Projections to 2017* (publication of the US Department of Agriculture, Economic Research Service, Washington, DC, February 2008).

34. US Department of Agriculture, *Agricultural Projections to 2020* (USDA publication, Interagency Agricultural Projections Committee, Long-term Projections Report OCE–2011–1, US Department of Agriculture, Washington, DC, February 2011); US Department of Agriculture, "Oil Crops Yearbook" (report, Economic Research Service, US Department of Agriculture, Washington, DC), www.ers.usda.gov/data-products/oil-crops-yearbook.aspx.

35. H. Huo, M. Wang, C. Boyd, and V. Putsche, *Life-Cycle Assessment of Energy and Greenhouse Gas Effects of Soybean-Derived Biodiesel and Renewable Fuels* (report no. ANL/ESD/08–2, Argonne National Laboratory, 2008).

36. B. H. Pro, R. Hammerschlag, and P. Mazza, "Energy and Land-Use Impacts of Sustainable Transportation Scenarios," *Journal of Cleaner Production* 13, nos. 13–14 (2005): 1309–19.

37. USDA, *Agricultural Projections to 2016*.

38. P. De Pous, *EEB Analysis of EU's Revised Biofuels and Bioenergy Policy* (publication of the European Environmental Bureau, Brussels, Belgium, March 19, 2009).

39. J. Hill, E. Nelson, and D. Tilman, "Environmental, Economic, and Energetic Costs and Benefits of Biodiesel and Ethanol Biofuels," *Proceedings of the National Academy of Sciences of the United States of America (PNAS)* 103, no. 30 (2006): 11206–10.

40. US Environmental Protection Agency, *Inventory of US Greenhouse Gas Emission and Sinks: 1990–2006* (EPA report no. 430-R-08-005, Washington, DC, April 2008).

41. USDA, *Agricultural Projections to 2017*.

42. Ibid.

43. D. Mitchell, "A Note on Rising Food Prices" (Policy Research Working Paper 4682, World Bank Development Prospects Group, July 2008), www-wds.worldbank.org/external /default/WDSContentServer/IW3P/IB/2008/07/28/000020439_20080728103002/Rendered /INDEX/WP4682.txt.

44. Ibid.

45. For greater detail about these mechanisms, see: International Union of Food Science and Technology, "Impacts of Biofuel Production on Food Security," IUFoST *Scientific Information Bulletin*, Washington, DC, March 2010.

46. L. H. Ziska, G. B. Runion, M. Tomecek, S. A. Prior, H. A. Torbet, and R. C. Sicher, "An Evaluation of Cassava, Sweet Potato and Field Corn as Potential Carbohydrate Sources for Bioethanol Production in Alabama and Maryland," *Biomass and Bioenergy* 33 (2009): 1503–8, www.cabi.org/Uploads/File/GlobalSummit/Lewis%20Ziska%20final%20paper.pdf.

47. This growth rate is based upon trends from 1960 to 2006. For more detail, see: US Department of Agriculture, *Agricultural Projections to 2016* (publication of US Department of Agriculture, Economic Research Service, Washington, DC, February 2007).

48. The emissions are estimated in the following fashion: *Emission rate (g/mill. Btu fuel at pumps) × Total fuel available at pumps (mill. Btu)*. For a detailed explanation for how emissions were estimated, see: McDonnell and Lin, "The Use of Biodiesel in Railways," 25.

Healthy, Oil-Free Transportation

10

The Role of Walking and Bicycling in Reducing Oil Dependence

KEVIN MILLS, JD[1]

In a country where many view bicycles as mere toys and walking as a relic of a bygone era, why would anyone posit that walking and bicycling (or "active transportation") are essential to a serious strategy to rein in America's oil dependence? The simple answer is that overreliance on motor vehicles is at the center of our oil dependence, and walking and bicycling are the most cost-effective ways to curb a substantial portion of projected growth in vehicle-miles traveled. A convergence of popular demand for safe and convenient places to walk or bicycle and fiscal constraints that compel policy makers to make the most of every tax dollar add up to a compelling case for increased investment in active transportation as an integral piece of our strategy to manage our nation's oil dependence.

At the root of this potential for active transportation to supplant certain driving trips is the surprising prevalence of short driving trips in America. Half of all trips taken in the United States are within a 20-minute bicycle ride (3 miles or less), and a quarter of overall trips are within a 20-minute walk (1 mile or less), and yet the vast majority of these short trips are taken by motor vehicle. This chapter will examine the potential to save oil by shifting a portion of these short driving trips to active transportation. Among the conclusions of this analysis is that 6.5 percent of projected oil consumption by all cars and light trucks could be avoided in the year 2050 by cost-effective investment in safe and convenient active-transportation systems, and that Americans could save over $900 billion at the pump between now and then at today's gas prices (not adjusted for inflation).

The Crux of the Oil Problem Is Insatiable Gasoline Demand

America cannot meaningfully reduce its oil dependence without curbing the demand for gasoline to power our cars and trucks. Over 70 percent of the oil used in the United States is used to fuel our transportation vehicles,[2] and most of this is for our personal automobiles.[3] Cars and light trucks have consumed a growing percentage of oil for decades as miles traveled have increased at a much faster rate than the US population.[4] This is due to more people driving more cars for more miles per car.[5]

The options for reining in this demand are to get more miles out of a gallon of gas, switch fuels, or drive fewer miles. Between 1981 and 2003, all technological efficiency gains were invested in increasing the size, weight, and power of vehicles rather than enabling them to go farther on a gallon of fuel. Vehicle-miles traveled grew relentlessly throughout this period, and oil demand for personal vehicles rose roughly proportionally to increased travel. Even though oil demand is expected to moderate in the future due to planned increases in Corporate Average Fuel Economy (CAFE) standards, growth in vehicle-miles will have to be managed to avoid losing the benefits of building cars and light trucks with improved fuel economy.

The history of efforts to switch to vehicle fuels other than oil has been similarly checkered, with little prospect of achieving meaningful reductions in overall oil demand in the short run and only limited potential in the longer run. The expense, logistical challenges, and environmental trade-offs inherent in changing our ubiquitous refueling infrastructure and developing a compatible vehicle fleet point to the wisdom of maintaining sober expectations. While efficiency gains and alternative fuels can help curb our appetite for oil, it would be foolish to put all our eggs in those baskets, given our track record of consistently failing to prioritize oil dependence when designing vehicles and the infrastructure for fueling them.

Over the past half century, increases in vehicle-miles traveled have been the engine of America's growing oil dependence. Throughout this period, increases in both vehicle-miles traveled and total fuel consumption have outpaced growth in our gross domestic product. In the face of sustained increases in gasoline prices and a stubborn recession, there has been a modest decrease in per-capita vehicle-miles traveled since 2005, but rising population or economic recovery may very well renew prior growth patterns in miles driven. Since population increases are projected and an economic recovery is desired and likely, substantial gains in fuel economy and in the use of alternative fuels could be lost to a contrary demand curve for travel.

Thus, an effective strategy for reducing oil consumption would be holistic, marrying gains in fuel economy and alternative fuels with efforts to manage vehicle-miles traveled. All the means of reducing oil dependence need to be on the table. Lacking a silver bullet, managing oil dependence is a matter of combining many interventions, each of which could reduce overall oil demand by 1–5 percent, according to David Greene, a research scientist with the University of Tennessee and Oak Ridge National Laboratories. Walking and bicycling could easily make a mid-range contribution in this context, according to the scenario calculations that appear later in this chapter. The 6.5 percent estimate from this analysis for reductions in demand from cars and light trucks in 2050 would correlate to a 2.6 percent reduction in overall oil consumption. Reductions of this magnitude would make an outsized contribution toward achieving a sustainable balance of supply and demand, relieving market pressures and vulnerabilities such as those that drove oil prices over $100 per barrel twice in the past few years.

Active Transportation Will Make a Critical Contribution to Managing Oil Demand

Active transportation, which includes walking and bicycling, as a tool to deploy in the effort to manage oil demand is particularly appealing for many reasons, including its cost-effectiveness and rich set of co-benefits (such as health), as well as the relatively short time frame in which bicycling and walking can begin to make an impact.

Communities across America are discovering that there is robust latent demand for safe and convenient places to walk and bicycle. Cities and towns that have created networks of active-transportation infrastructure that connect residences to workplaces, schools, shops, recreation, and transit have experienced substantial increases in utilitarian trips taken on foot or by bicycle. Despite conventional wisdom that Americans won't leave their cars at home, there is significant and growing evidence that active transportation is like the Field of Dreams: if you build it (i.e., create functional active-transportation systems), they will come.

One set of evidence of this trend comes from an experiment set up by the United States Congress in the federal transportation bill passed in 2005. The Non-motorized Pilot Program designated four communities to dedicate $25 million each to filling gaps in their active-transportation networks and implementing complementary programs in order to see if these investments would shift some driving trips to walking or bicycling. The communities selected represented diverse geographic areas: Minneapolis, Minnesota, is a sizable city; Marin County, California, is a suburban jurisdiction; Columbia, Missouri, is a medium-sized college town; and Sheboygan County, Wisconsin, is made up of small towns and rural areas. Further, their experiences with

active transportation varied, with Sheboygan County having done very little to foster active transportation prior the pilot program. Despite the differences between the
pilot communities, they all experienced substantial growth in walking and bicycling
rates when they made investments in active-transportation infrastructure. Overall
across the pilots, bicycling increased 49 percent and walking increased 22 percent in
just three years (2007–2010),[6] resulting in a reduction of 37.8 million miles of avoided
driving in those four communities.[7] All four continue to work to build on this legacy,
and further dramatic increases in vehicle-miles avoided are expected.

The pilot communities are certainly not alone. Cities from Portland, Oregon, to
Washington, DC, and New York City have undergone striking transformations in
which safe and convenient facilities have translated into exponential growth in bicycling. The desire to switch to active transportation can be especially strong in densely
settled areas where it is particularly slow and expensive to use an automobile. In cities
with traffic congestion and steep parking rates, many people rely on active transportation as simply the most convenient and affordable way to get around. Convenience is
a factor even in the modest-sized city of Billings, Montana (population 100,000). The
Sneakers, Spokes, and Sparkplug Challenge pits bicyclists, pedestrians, and drivers
against one another in completing a set of tasks around town. The bicyclists always
win, with the pedestrians often finishing before the drivers. Further, the Chamber of
Commerce sees Billings' trail system as essential to attracting talented individuals to
live and work there.

The trend is not limited to cities. Small towns and rural areas, on average, experience walking and bicycling rates only modestly lower than a typical city and greater
than many suburbs.[8] Walking and bicycling trails have brought economic revitalization on a broad scale to many small towns, like Cumberland and Frostburg, Maryland. Making it safe to walk and bicycle is also vital to maintaining a family-friendly
environment, and creating the option to save on gasoline can be a lifeline for those on
limited incomes.

Investments in utilitarian bicycling and walking systems can make a significant
contribution to reducing oil demand primarily because of the surprising prevalence of
short driving trips in America. Of all trips in the United States, 48 percent are within
three miles or less and 27 percent are within a mile, and yet 72 percent of these short
trips are taken by motor vehicle. Many experienced bicyclists and pedestrians exceed
these distances on a routine basis but, with 98 percent of walking trips and 85 percent of bicycle trips covering three miles or less, the biggest opportunity to make a
sizable dent in oil use (and, incidentally, obesity rates) is to increase the share of the

population that make some short trips using their own power. This is further under-scored by the fact that the national share of all trips of three miles or less that are taken on foot or by bicycle (22.8 percent) is nearly double the share of trips of all lengths (11.9 percent).

Because of the significance of short trips, the calculations in this chapter are based solely on replacing a portion of short car trips by a growing number of active travel-ers. This mainstream mass of willing but cautious pedestrians and bicyclists require safe and convenient routes to routine destinations in order to feel comfortable enough to take part. The calculations disregard longer trips altogether in order to dramatize the potency of a strategy focused on short trips.

Calculating the Oil Savings from Active Transportation

Calculating the oil savings that could be achieved by active transportation in 2050 must begin with assumptions about the world nearly 40 years in the future. The base assumption used here is that there will be just over a doubling of the share of trips taken by bicycling and walking over the 40-year period from 2010 to 2050, with a straight-line year-over-year increase during that period. This rate of increase (0.325 percent per year) is significantly below the growth rate in active transportation be-tween 2003 and 2009 (0.4 percent), according to the National Household Travel Sur-vey (conducted by the US Department of Transportation).

To help the reader imagine how this projected level of active transportation might look in the American context, picture Minneapolis today, where 28 percent of all trips involve walking or bicycling (19.6 percent solely by active transportation plus 8.5 per-cent of trips via transit accessed by walking or bicycling). If over the course of the next 40 years the nation were to steadily approach the rates currently experienced in Minneapolis, America could exceed the oil savings and other benefits outlined here.

Under this substantial but conservative scenario, which assumes that only short trips will shift to walking or bicycling, the percentage of total trips in America made by walking or bicycling would increase from 11.9 percent to 25 percent by the year 2050.[9] Such a shift would mean nearly 200 billion driving miles avoided due to active transportation annually by 2050.[10] This shift would increase the amount of oil saved by active transportation in the United States to 245 million barrels of oil (over 10 bil-lion gallons) in the year 2050 alone, assuming current levels of fuel economy. Fuel economy increases could reduce this calculation, but active-transportation trips of greater than three miles would increase it.

This level of savings equates to about 6.5 percent of the Energy Information Ad-ministration's projection for total oil consumption by US cars and light trucks in the year 2050. Assuming that personal vehicles continue to account for 40 percent of total oil consumption, active transportation under this scenario would save between 2.5 percent and 3.0 percent of total US oil demand in 2050. This puts active transportation squarely in the mid-range of substantial solutions sought to manage our nation's oil dependence.

Using the average gas price of $3.70 per gallon at the time of this writing, the oil conserved in 2050 would generate estimated direct savings at the gas pump of $38 billion (in today's dollars). The cumulative picture is even more impressive. Assuming a linear rate of diversion of vehicle-miles traveled between 2010 and 2050, the total amount of oil conserved by active transportation over that period would be 5.87 billion barrels (or about 246 billion gallons), equivalent to $912 billion at today's gas prices. Especially since so much of the money spent on gasoline ends up overseas, individuals and the nation's economy would benefit from greater disposable income to spend closer to home.

These calculations do not count many sources of additional savings such as reduced traffic injuries and fatalities, decreased air and climate pollution, a healthier and more active population, greater access to employment for low-income groups, reduced wear and tear on roads, and diminished need to expand road capacity. These are among the many co-benefits that flow from investment in active transportation.

Rich Set of Co-Benefits

The best reason to invest in increasing the share of trips taken via bicycling and walking, though, may be that active transportation mitigates many societal problems all at once. Reducing our dependence on oil is an important objective for policy makers, but given a plethora of social challenges we need policies that address oil use and other objectives simultaneously. In an era of fiscal constraint, we simply cannot afford to solve one problem while creating or ignoring others. Transportation policies of the past 50 years, with a single-minded focus on building highways to move cars faster, has created undeniable benefits but also unanticipated problems beyond oil dependence, such as sprawling land use, air pollution, and traffic fatalities. Just as a prudent investor diversifies his or her holdings in order to minimize risk and maximize return, or a farmer plants different crops to maintain the productivity of his or her land, a holistic policy of fostering a balanced mix of transportation options helps to make livable communities and maximize public benefits. Given the rich set of co-benefits, investments in active transportation represent a cost-effective "no regrets" response to oil dependence.

Major categories of co-benefits of active transportation are public health, the environment and the economy. Public health is advanced by integrating physical activity into the daily lives of a populace suffering from an obesity epidemic and many chronic diseases associated with sedentary lifestyles. If 25 percent of trips were to be completed by active transportation in 2050, as much as $28 billion per year in health-care cost savings could be realized (with no adjustment for inflation). These savings result from short walking and bicycling trips that generate a quantity of physical activity equivalent to every American exercising an average of nine minutes per day. Our assumption is that half of those minutes will come from people who do not currently meet the Centers for Disease Control's recommendation of 30 minutes of moderate exercise on most days, and half who do. We did not attribute any economic benefit to the latter group, even though there undoubtedly are some. We further assumed modest speeds of 3 miles per hour for walking and 10 miles per hour for bicycling, and that half of the trips would be completed by each of these means.

The value of CO_2 reductions alone from short bicycling and walking trips between 2010 and 2050, based on the same assumptions as above, would generate another $38 billion in cumulative savings at current trading prices on the European carbon market.[11] Further, the advantages of shifting short driving trips to non-polluting forms of transportation is more marked in the case of smog precursors because these pollutants are emitted at very elevated levels during the first few miles of any driving trip. Studies have suggested that the vast majority of hydrocarbons, for instance, are emitted in the first two minutes of vehicle operation.

The economic benefit of these co-benefits is substantial. In 2008, I co-authored a report with Thomas Gotschi—*Active Transportation for America*—which quantified the benefits of investing in shifting short driving trips to bicycling and walking. Using data available at the time, we calculated status quo known benefits of more than $4 billion per year. However, since that time new federal data on the share of trips taken via walking and bicycling indicate that in the intervening six years between data sets, active transportation had increased to the point where we are already approaching our modest scenario in which we projected annual savings due to active transportation to exceed $10 billion per year (assuming $3 per gallon for gas).[12] Under our substantial mode-shift scenario, where 25 percent of trips would be made by bicycling or walking short distances, the economic benefits could rise to $65.9 billion annually (without adjusting for inflation).[13] This figure is based on estimates of fuel savings (assuming $4 per gallon for gas), reduced CO_2 emissions, and the health-care savings of increased physical activity.

These calculations do not account for many other benefits, including time saved in reducing traffic congestion and average commuting times, environmental and health benefits stemming from reduction in air pollution from pollutants other than CO_2, increased safety from avoided death and injury, economic stimulus from construction, rises in local commercial and residential real estate values,[14] avoided automobile depreciation and operational expenses such as insurance and maintenance, reduced spending on road maintenance and increases in road capacity, health benefits among those who are already active, and gains in the quality of life. Trail systems and walkable neighborhoods are increasingly seen as a magnet for talented workers. Further, active-transportation investments improve access to transit, ensuring that we get more use out of our public transportation facilities, which account for nearly 20 percent of overall surface-transportation expenditures.

Bicycling and walking are also a metaphor for smaller, simpler government. With concerns about the federal budget deficit running high, it is telling to calculate the savings to the federal budget and US taxpayers from strategic investment in active-transportation networks. Even from this narrower perspective, which ignores broader societal benefits, active-transportation investment is highly cost-effective, returning more than $4 to the federal Treasury for each dollar invested, in just the first decade. This return would multiply further over time. This is accomplished in three ways: (1) savings on road construction and maintenance by providing a much cheaper way to move the same number of people for some short trips, (2) health-care and safety savings due to increased physical activity, cleaner air, and fewer auto crashes,[15] and (3) the creation of jobs needed to construct active-transportation facilities, which creates more jobs per dollar than building roads, and leads to additional local economic development.

Moving Forward

By 2050, relatively modest strategic investments to complete active-transportation networks could save 10 billion gallons of oil per year, equivalent to 6.5 percent of projected oil consumption by cars and light trucks, and enjoy about $66 billion each year in health, economic, and environmental benefits. Between now and then, Americans could save over $900 billion at the gas pump.

What would it take to realize these benefits along with a rich set of co-benefits? The top two reasons Americans give for not walking or bicycling more are the lack of safe and convenient infrastructure, and land-use patterns. Where the distances are accessible and the conditions safe, which typically means separation from automobile traffic for most novice and intermediate bicyclists, Americans will travel under their own power at increased rates.

Focused investment to fill gaps in active-transportation networks is a cost-effective prescription for increasing active-transportation mode share to unlock these benefits. A sidewalk or trail that simply ends short of a destination represents an opportunity to make valuable connections on the cheap by building on preexisting infrastructure in order to create utilitarian networks.

While walking and bicycling already save money, improve health, and help reduce our nation's oil dependence, the unrealized potential for them to contribute more is vast. Already America is on the cusp of a shift in which active-transportation infrastructure is increasingly recognized as essential to any desirable community.

A federal policy framework designed to realize the oil savings potential of active transportation as well as health, safety, and environmental benefits would include the following:

- Continued dedicated funding for trails, bicycling, and walking through proven, cost-effective programs. The Transportation Enhancements program has been the lifeblood sustaining the active-transportation progress made over the past 20 years. Safe Routes to School and the Recreational Trails Program are also important core programs. These critical programs are wildly popular in communities across America and boast an impressive track record of success in improving the nation's transportation system for less than 2 percent of surface-transportation funding, but nonetheless they have come under intense fire in an increasingly divided Congress. These programs—which were consolidated into the Transportation Alternatives program by a new federal transportation bill passed in the summer of 2012—deserve greater support in times of fiscal constraint because they deliver superior benefits for each taxpayer dollar than other options for managing short trips.
- New policies and initiatives focused on creating active-transportation systems. Concentrated investments to fill gaps in systems are needed in order to create safe routes to everywhere. The key to achieving the oil savings detailed in this chapter is to make walking and bicycling safe and convenient by connecting homes, businesses, schools, and other routine destinations with networks of active-transportation infrastructure that Americans of all ages and abilities will be comfortable using.

Notes

1. The author acknowledges the invaluable assistance of his intern, Isaac Binkovitz, in researching and preparing this chapter.

2. See: www.eia.gov/emeu/aer/pdf/pecss_diagram_2009.pdf.

3. Portion of total oil used for personal cars and trucks: 40.7 percent (which is 56 percent of all oil used for transportation; see: www.rightofway.org/research/newoilage.pdf at 2, 10.

4. Rails to Trails Conservancy, *Active Transportation for America: The Case for Increased Federal Investment in Bicycling and Walking* (RTC publication, 2008), 21, www.railstotrails.org/resources/documents/whatwedo/atfa/ATFA_20081020.pdf.

5. J. DiCicco and F. An, *Automakers' Corporate Carbon Burdens: Reframing Public Policy on Automobiles, Oil and Climate* (publication of Environmental Defense, 2002), cleartheair.edf.org/documents/2220_AutomakersCorporateCarbonBurdens.pdf.

6. Congressional briefing on the results of the Non-motorized Pilot Program, May 12, 2011. Figures represent a weighted average by population.

7. See, for example, this Federal Highway Administration report to Congress, released April 2012: www.fhwa.dot.gov/environment/bicycle_pedestrian/ntpp/2012_report/.

8. Beyond Urban Centers report, published in January 2012; see: www.railstotrails.org/ourWork/reports/beyondurbancenters.html.

9. The 11.9 percent figure is the actual figure for 2009, according to the National Household Travel Survey. For the sake of simplicity, it is used as a year 2010 baseline from which increases are calculated. (See: nhts.ornl.gov/.)

10. This is calculated using Energy Information Administration projections for 2050 oil demand of 3.8 billion barrels annually, or 10.4 million barrels per day, for cars and light trucks and dividing by the share of projected vehicle-miles traveled that will be accomplished by bicycling and walking at the assumed 2050 mode share.

11. On August 21, 2011: 12.5 Euros per metric ton of carbon, which equals $17.98 per metric ton of carbon.

12. National Household Travel Survey, 2009 (publication of the Federal Highway Administration, US Department of Transportation, Washington, DC).

13. See: Rails to Trails Conservancy, *Active Transportation for America*; page 38 details the underlying assumptions and calculations for the monetary value of benefits from bicycling and walking.

14. C. Leinberger and M. Alfonzo, *Walk this Way: The Economic Promise of Walkable Places in Metropolitan Washington, D.C.* (publication of the Metropolitan Policy Program, Brookings Institution, 2012), www.brookings.edu/research/papers/2012/05/25-walkable-places-leinberger.

15. The federal government pays 30 percent of the nation's medical bills through Medicare, Medicaid, and federal and military insurance.

11 Building an Optimized Freight Transportation System

ALAN S. DRAKE

The United States of America created a transportation system for both passengers and freight that was based upon a reality that no longer exists. Cheap, abundant, and largely domestic oil as well as an endless supply of land for expansion were all assumed. Fuel taxes too low to pay for maintenance and eventual replacement of the roads were levied, and we assumed that point-to-point trucking over "free"-ways was the most efficient way to deliver goods—efficiency being measured solely by time and flexibility.

Today, reality is quite different. Oil is no longer cheap, abundant, or largely domestic. It is reasonable to expect that the future will be even more oil-constrained. The underfunding of maintenance and the replacement of highways is becoming critical. "Free"-ways are not so free in the long run—and the "long run" has finally arrived. Climate change is becoming an increasing concern.

Yet we are working with a freight transportation system designed in the 1950s and largely completed by the 1970s. When federal regulation ended in 1980, the freight system had railroads carrying low-value, time-insensitive cargoes like coal, lumber, and gravel. Trucks dominated the higher value cargoes. Since deregulation, rail has captured 15 percent of the fresh vegetable market and up to 95 percent of containers on specific routes, and 0 percent on other routes.

Today, oil-powered railroads carry slightly over half the intercity ton-miles of trucks, but trucks strongly dominate the value of cargoes moved—and the freight revenue. Trucks may receive five times the revenue for slightly fewer ton-miles.

How would we design a freight transportation system optimized for the world of today? And, looking forward, what would we design for the foreseeable future, so that our newly designed transportation system would not be functionally obsolete the day that we completed it?

Considering a future with economic, energy, and environmental constraints growing in importance, a new freight transportation system would:

- Transport freight with a fuel other than oil, and ideally with a renewable fuel.
- Be extremely efficient, in terms of energy, labor, and materials, to operate.
- Be cheaper to build as a new system than it would be to maintain the old system.
- Be cost-efficient, having a positive Elasticity of Supply: the marginal supply of additional transportation costs less than the average cost, so that the more it is used, the less it costs per unit (see figs. 11.1 and 11.2).
- Require minimal additional taking of land.
- Be faster and more reliable than our existing transportation system.

The backbone of a future freight system should be electrified, expanded, and improved railroads, with trucking continuing to serve for shorter distances, "last mile" service, and local deliveries, and with a corresponding shift in freight revenues from trucks to rail and, overall, significantly less money and resources spent on moving freight. Long- and even medium-distance freight should be moved by electrified rail with double-stack clearances—except when even improved and expanded rail service cannot meet the shipper's needs. Such a system has the potential to fulfill all six of the criteria above, with matching the speed and reliability of trucking being the most challenging. The role of trucking would diminish, but not disappear.

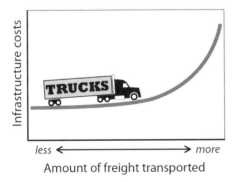

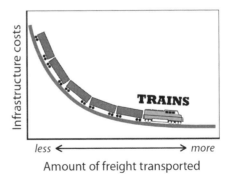

Figure 11.1   Efficiency curve of trucks as the amount of freight transported increases.

Figure 11.2   Efficiency curve of trains as the amount of freight transported increases.

Inland and coastal maritime shipping would have a place in the future as well due to the exceptional fuel efficiency of barges. The Atlantic Coast Intercoastal canal system is lightly used and has significant potential for growth. A recent example is CP Rail taking Bakken oil from North Dakota to the Hudson River by rail, where the oil was barged down to a Philadelphia refinery.

There are discussions of creating a system of scheduled container barges moving up and down the Mississippi River. Reduced ice due to climate change is extending the Great Lakes shipping season, and rising fuel costs are making ship transport more attractive. However, in a very high-oil-price environment, electrified trains would be more desirable and would likely stifle increased barge traffic growth.

While trains use more BTUs to move a ton-mile than do ships, if the train BTUs (electricity) are significantly cheaper than ship BTUs (oil), rail will have a fuel-cost advantage over oil-burning ships. The privately owned US railroads of today see the immediate potential and are investing an astounding 18 percent of their gross revenues in capital projects[1]—but this is not enough to adequately capitalize on today's opportunities or tomorrow's potential.

It is worth noting that although railroads move slightly more ton-miles than do intercity heavy trucks, the total revenues for all major (Class I) US railroads—$63.8 billion for 2011[2]—is less than the cost of the 700 million barrels of refined diesel that intercity heavy trucks will burn in 2012.[3] Add to this the annual road maintenance subsidy of trucking of $30 to $40 billion every year.[4]

Transferring current truck freight, and much of the associated revenue, to an electrified and expanded rail system will easily pay for the required investment to do so, due to the resulting reductions in highway maintenance costs, administration, and related expenses, as well as reduced fatalities and pollution impacts.[5] But the investment in rail has to be made first—and the railroads cannot be expected to invest much more than 18 percent of their current gross revenues. The only other corporations where 18 percent of gross revenues are reinvested in the business are the hottest high-tech firms. One percent to 5 percent of gross revenues going into capital expenditures is much more typical of American business.

Public policy should concern itself with creating incentives and public-private partnerships that more clearly align National Security, macro-economic, social, and environmental goals with the goals of the privately owned railroads. Society is concerned with a variety of goals, such as delivering food and critical materials during a severe and prolonged oil-supply shortfall. These are goals that railroads view as externalities—however positive they may be for the rest of society.

Investments for Today

There are a number of basic investments in moving freight by electrified and expanded railroads that will benefit our economy, the environment, and society overall. The sooner that these investments are made, the better it will be for our nation.

Electrify 35,000 Miles of Main Rail Lines in the United States

The United States transitioned from steam locomotives to diesel-electric locomotives, and not to a mixture of electric and diesel-fueled locomotives, reportedly because General Motors threatened to blackball any railroad that electrified after World War II. Nevertheless, rail electrification is a very well-proven technology, and building and operating electrified rail lines in virtually any environment has over a century of experience to draw upon. The Russians finished electrifying the Trans-Siberian Railroad in 2002 and electrified the line up to the Arctic port of Murmansk in 2005. China is electrifying 12,400 miles of their rail lines this decade, and the Indians are building 1,675 miles of all new double-stack electrified freight rail lines. England and Wales have recently announced a £9.4 billion package of rail investments, with electrification of existing rail lines and new electrified rail lines taking the bulk of the investment. All major US rail lines—except for Amtrak's Northeast Corridor and some commuter lines—are freight lines first, and the Northeast Corridor and Long Island Railroad are already largely electrified, and Caltrains soon will be.

Electrifying the 100,000 miles of branch lines in the United States is not as critical as electrifying the main lines. The bulk of the ton-miles are on the main lines, and electrification of these will capture a large majority of the benefits. The minimum traffic required for electrification in the United States can be determined at a later date. In Europe, lines with as few as six trains per day are considered to be worth electrifying.

An expedited build-out of this new infrastructure could electrify 35,000 miles in seven years (six years per schedule and add one year for Murphy's Law). The level of effort required would be comparable to the effort currently devoted to boiling more tar out of the sands in northern Alberta. Britain has ordered a work train that can electrify a mile of track in a single shift. Using standardized costs, it would take about $80 billion to electrify 35,000 miles of US railroads.[6] However, Matt Rose, CEO of BNSF, has stated that the main lines of BNSF Railroad could be electrified for $10 billion.[7] Extrapolating that to the rest of the United States produces a cost estimate close to $50 billion. By comparison, current rail electrification cost estimates of Indian Railways applied to 35,000 miles result in an estimate closer to $35 billion, although Indian wages are far lower than American wages.

Operating 80 percent of the existing rail ton-miles and half of the current trucking ton-miles on electrified main lines would increase current US electrical demand by approximately 1.6 percent.[8] One requirement of electrification—bringing high-voltage transmission along the rail right-of-way (ROW) in remote areas—can also be a tremendous asset. American financier T. Boone Pickens once proposed electrifying a BNSF rail line if he could use the rail ROW to transmit his wind-generated electricity to market.[9] Unfortunately, Mr. Pickens never followed through and built his West Texas wind farms, but the basic concept is valid. Electrified railroads can also serve as transmission corridors for renewable electricity.

Electrification expands track capacity by roughly 15 percent in Europe due to faster braking and acceleration.[10] No one is sure about the amount of increase in track capacity with North American railroads, but it would be substantial. The president of Norfolk-Southern has speculated publicly about electrifying their Heartland Corridor just to increase track capacity, with any fuel savings and increased locomotive reliability being a bonus.[11]

A number of years ago, Matt Rose, CEO of BNSF, in answer to a shareholder question at the annual meeting, said that rail electrification was not cost-justified at the then price of oil—$60 per barrel—but it certainly would be at $150 per barrel. That day may not be far distant. However, at the most recent shareholder meeting of Berkshire Hathaway (now owner of BNSF), Matt Rose answered a shareholder question about electrification by saying, "We will do LNG (Liquified Natural Gas) first. Electrification will have to wait for $7 or $8 gallon diesel."

A recent IMF (International Monetary Fund) study suggested that real (inflation adjusted) oil prices will double in the next decade, which is close to the time required to electrify our rail mainlines.[12] And a doubling of the price of oil will indeed bring diesel to $7–8 per gallon.

Electric locomotives are about 2.5 times as efficient (BTU electricity to BTU diesel) as their diesel-electric counterparts on flat rural tracks. However, in mountains and in developed areas, electric locomotives are about three times as efficient.[13] The difference is that when an electric locomotive brakes (as it must in mountains and urban areas), it turns the motor into a generator and feeds power back into the line.

Furthermore, electric locomotives are more than twice as reliable as the diesel-electric locomotives used today in North America—a distinct operational advantage for busy rail lines. After 2015, new US diesel-electric locomotives will have to meet much stricter pollution requirements, which will add to their operational costs and reduce their reliability. Such new pollution requirements will, of course, not affect electric locomotives.

Build Intermodal Terminals

Intermodal terminals transfer cargo—typically in containers—from one transportation mode to another. In common usage, *intermodal terminals* has the more limited meaning of transfers between truck and rail, although ports have long been "intermodal terminals" serving truck, ship, and rail. Much of American manufacturing and most of the distribution is not set up for direct rail service at the loading dock. Trucks are often used as the "last mile" service, with rail providing service to a nearby intermodal terminal, despite the additional cost of handling the intermodal transfer. This "last mile" service dramatically shifts the rail-to-truck ton-mile ratio toward rail.

The railroads want the economies of scale and the revenues that come with large terminals. New intermodal terminals are being built at a furious pace, though they are usually dedicated to a single railroad. The Norfolk-Southern intermodal terminal in Columbus, Ohio, for example, receives four container trains a day from Chicago (usually transferred from West Coast trains) and two from Norfolk, Virginia. Between 200 and 250 containers on each train are sorted for either transfer to trucks or north- and southbound trains, or they are routed farther east and west.

Relatively few intermodal terminals can achieve such volumes. Limiting intermodal terminals to high-volume locations means relatively few will be needed, and many shippers find that they "can't get there from here" and resort to more-expensive truck-based transport. Limited numbers of intermodal terminals also means that the "last mile" on trucks is longer than it would be if there were more terminals.

There are alternative paradigms. Florida East Coast Railroad (FEC) has five intermodal centers on just 317 miles of mainline track. Shippers use them not just for interstate shipping, but also for shipping from one end of Florida to the other. The tightly scheduled service has increased the rail modal share (80 percent of FEC's revenues are from intermodal containers), and has gradually decreased the truck ton-miles on Florida highways. In Chippewa, Wisconsin, CN Railroad has recently opened up a mini–intermodal terminal, with one intermodal train stopping there two days per week.

As freight brokers find that they can indeed ship by rail (a much cheaper option), the volumes at intermodal terminals will increase. Higher volumes will justify a greater density of intermodal terminals. This is the virtuous positive feedback cycle that public policy should accelerate. At the very least, rail-to-truck transfers should be accommodated as often as possible. In the distant past, virtually every factory and warehouse had a rail loading dock. Tracks can still be seen embedded in city streets here and there. The proportion of warehouses served by rail steadily declined for many decades, but this trend has recently reversed. For example, Toyota demanded that the state government of Texas build an eight-mile rail spur to BNSF from a proposed San Antonio plant so that Toyota could ship by either Union Pacific or BNSF.

Improve Double-Stack Container Clearances

Containers either international (8' wide × 8' high × 20' or 40' long) or domestic (8' wide, 8' or 9.5' high × 53' long), can be shipped on all North American rail lines, one container per well car (that is, a container-carrying railcar). Many rail lines lack the clearances required for double-stack containers—one container stacked on top another.

However, it is much more efficient to stack containers two high on each well car. Twice as many containers can be shipped on a train of a given length. A container well car can carry one or two containers; the second container adds no extra tare, or non-payload, weight. The aerodynamic drag of a second container is much less than that of the first one plus the railcar. Double-stack container trains are extraordinarily energy-, labor-, and logistically efficient.

Many bridges and tunnels were built before the era of double-stack container trains, and they do not provide adequate clearances. The railroads have been diligently increasing these clearances on most busy rail lines, often with partial funding from state departments of transportation. Recently, Norfolk-Southern transformed a major coal line into a high-volume container line between Chicago and Norfolk, Virginia, with partial funding from several state DOTs. However, the task is far from complete, especially in the Northeast. But once done, clearances can be maintained indefinitely with minimal effort and cost. They are a permanent investment in transportation efficiency.

Add Double Tracks

Many busy rail lines operate like a single-lane street with two-way traffic. An eastbound train has to pull over onto a siding and wait for all the westbound trains to clear the single track before proceeding. This obviously slows down shipments, adds to labor and rolling-stock capital costs, and reduces the reliability of delivering cargoes on time. Adding a second track can more than double a line's capacity while speeding up shipments. Quicker shipments mean less capital is needed for locomotives and railcars, and this capital savings can be applied to the capital cost of adding a second track.

Often a second track was removed decades ago as subsidized highways took much of the rail business away. In almost all cases, the rail ROW is still wide enough to add a second track without requiring additional land. Today, about 18,000 miles of US railroads are double- or triple-track. Adding an extra 12,000–15,000 miles of a second track would significantly increase our railroad capacity, speed up shipments, and lower operating costs. At $7 million per mile (for simple new tracks),[14] that would cost $84–105 billion. However, double-tracking bridges, tunnels, and other difficult places would certainly increase that estimate.

Build Rail-over-Rail Bridges

Just as a road overpass increases traffic speed and road capacity over a stoplight intersection, so does a rail-over-rail bridge. Rail-over-rail bridges are needed to improve the speed and reliability of rail freight as well as removing capacity-limiting bottlenecks.

Just one such rail-over-rail bridge, built in Kansas City a few years ago at a crucial bottleneck, had positive ripple effects on train scheduling across much of the United States. The $3 billion public-private partnership CREATE relies heavily on such rail overpasses to unclog the major rail bottlenecks of Chicago. Nationwide, additional rail-over-rail bridges would not be much more than double the number needed in the Chicago area.

Improve Signals

All Class I railroads are required to install Positive Train Controls by 2015 on most of their lines—70,000–80,000 miles of track—at a 20-year cost of $10–14 billion.[15] Safety will be improved and track capacity increased, but the cost benefits are questionable compared to other rail investment options.

It is worth noting that freight trains killed 560 people in 2010, and trucks killed almost 5,000.[16] The greatest safety program is shifting freight from trucks to rail, but additional research is needed to quantify these safety improvements.

Allow for Increased Speeds with Grade Separation

At-grade crossings of roads and railroads are the major source of accidents, and they can hinder rail operations. In most cases, the railroads were there first, and common law suggests that the more recent infrastructure, roads, should pay so as to not interfere with the grandfathered railroads. Without grade separation, with even the best-quality track, passenger service is limited to 89 mph and freight to 80 mph under FRA regulations.[17]

Revisit Rolling Highways and Trailer on Flat Car

Before the widespread market acceptance of containers, an earlier mode of intermodal service was to park truck trailer on railroad flat cars—Trailer on Flat Car. This has significantly lower efficiency (4× as energy efficient as a truck, compared to 9× for double stack containers),[18] but the handling and infrastructure required is much less.

The overall tonnage handled with Trailer on Flat Car has held steady or decreased each year as rail container traffic grows rapidly, according to each new statistical report.[19]

The efficiency gains and fuel shift from diesel to rail electrification should make Trailer on Flat Car more economically attractive, as would reducing or eliminating the annual road-maintenance subsidy to truckers. Trailer on Flat Car transports just the truck trailer; a new tractor and driver pick up the trailer on the other end. This is more efficient, but less attractive to independent and other truck operators.

In Europe, a popular service is the "Rolling Highway." Truck operators park both their tractors and Trailers on Flat Cars and then ride in sleeper rail cars, getting their required rest, or ride in regular coaches for shorter hauls. No US railroad offers anything comparable, and it appears that new legislation would be required to authorize this service. However, it appears to have potential on several routes, such as I–81 and the Alaskan Railroad. The energy-efficiency gains are more modest (though still in excess of 2×, with domestic electricity replacing imported oil), but it would allow for rail capture of a higher percentage of the trucking market. Increasing fuel taxes on trucks enough to remove the road maintenance subsidy would make Rolling Highways more attractive.

Move to Semi–High Speed Rail for Passengers and Freight

True high-speed rail (HSR) in Europe and Asia is used exclusively for passengers and low-density freight like letters and packages. The physical structure of the tracks does not easily allow 180–220 mph passenger service and even medium-density freight to share the same tracks. The primary issue is in the curves and the tilt, or "super elevation," of the outside rail above the inside rail plus the radius of the turn. A secondary issue is that many freight trains have higher centers of gravity than passenger trains and higher axle loadings as well and cannot climb steep grades as well as HSR passenger trains.

The Swedish and Italian "tilt" trains are semi–high speed rail trains that tilt the passenger train to allow higher speeds on freight rail tracks. Unfortunately, an attempt to include "tilt" technology on the Amtrak Acela trains was generally considered a failure, and the technology appears problematic with American railroad regulations. However, there is a "sweet spot" where slightly slower passenger trains (100–150 mph) can share tracks with express freight (90–110 mph) carrying medium-density freight like fruits and vegetables, electronics, and fish, as well as letters and packages. SBB (Swiss Rail) will soon be offering 100 mph express freight on the same tracks with 150 mph passenger trains.

Conclusion

The transportation system we have today is in the process of failing. Financing road repairs, that is, continued trucking subsidies, is a growing conundrum in statehouses across the nation. "Fuel-cost adjustments" on lading bills impact almost every sector of the economy.[20] And the toll from air pollution and car-truck collisions continues to mount. We need a new, and better, freight transportation system, one that can stimulate the economy with lower and more stable costs for the transportation of goods. This chapter outlines the elements of what that new system should be for most intercity freight.

The privately owned railroads are making extraordinary levels of capital investment. They will, no doubt, get excellent returns for their investments. However, from a public policy perspective this self-financing, "lifting themselves by their own bootstraps" effort is far too little and far too slow. The imbalance is evident when considering:

1. the annual public subsidy for trucking—$30 to $40 billion;
2. the transfer of General Fund and other unrelated tax monies to subsidize roads and highways—in 2010, $80.2 billion plus net new borrowing by highway funds, and another $20.8 billion in direct subsidies to roads and highways, for a total of $101 billion in 2010 ($326.99 per capita);
3. the trucking fuel bill in 2012—likely $120+ billion;
4. the gross revenues for railroads—$63.8 billion for the five US Class I railroads in 2011; and
5. the planned 2012 capital investments by railroads—$13 billion.[21]

As a nation, we would be far better served to spend more billions on domestic capital investments, many of which will last a half century or longer, and many fewer tens of billions on importing foreign oil and subsidizing trucking via road and highway maintenance. There is a critical National Security need to create an oil-free transportation backbone in this country. Electrified railroads are the key component in creating this. In addition, there are significant economic, environmental, and employment benefits from electrifying and expanding our railroads.

I suggest a 33 percent investment tax credit for rail capital projects that increase track capacity or increase average track speed. Furthermore, I suggest a 50 percent investment tax credit on rail electrification, including electric locomotives, that starts to phase down to 33 percent after seven or eight years. Further, the subsidy to trucking should be phased out. Fuel taxes on diesel, natural gas, or other fuels should be raised to reflect the cost to repair the highways damaged by trucking. This is the old conservative principle that the "user pays."

The benefits of a new, and better, freight transportation system are both quite large and quite varied. The sooner we turn away from an old 1950s vision that no longer works toward a sustainable and efficient freight transportation system, the brighter our future will be.

Notes

1. This figure is based on SEC-reported gross revenues detailed in the American Association of Railroads press release "Freight Railroads Expect to Spend a Record $13 Billion on Capital Expenditures, Hire More than 15,000 in 2012" January 30, 2012; see: www.aar.org /NewsAndEvents/Press-Releases/2012/01/30-Investment-And-Hiring.aspx.

2. Data source: 2011 Revenues for BNSF, UP, CSX, N-S, and KCS Railroads; see: finance. yahoo.com.

3. Energy Information Administration, "AEO 2009 National Energy Modeling System: An Updated Reference Case Reflecting Provisions of the American Recovery and Reinvestment Act" (EIA report, US Department of Energy, Washington, DC, 2009); see table 46: "Transportation Sector Energy Use by Fuel Type Within a Mode." (Adjusted for decreased fuel-tax revenues.)

4. Federal Highway Administration, Highway Statistics 2009 (FHWA annual report, US Department of Transportation, Washington, DC, 2012); see table HF–10 "Funding for Highways and Disposition of Highway-User Revenues, All Units of Government, 2010," www .fhwa.dot.gov/policyinformation/statistics/2010/hf10.cfm. This is a modest estimate; other sources calculate the subsidy at closer to $60 billion per year; see, for example: truecostblog. com/2009/06/02/the-hidden-trucking-industry-subsidy. However, this calculation does not account for weather damage, which varies significantly by state.

5. According to the Federal Highway Administration, roads and highways received $61.6 billion from General Fund transfers, property taxes, and other non-highway sources in 2009 (www.fhwa.dot.gov/policyinformation/statistics/ 2009/hf10.cfm). Heavy trucks paid about $9 billion in fuel taxes. Besides new and rebuilt roads, highway patrols, administration, and other related expenses ($127.5 billion in 2009)—some of which should be apportioned to trucking—$49.4 billion was spent on road maintenance. Road damage is proportional to the fourth power of the axle weight; this means that heavy trucks are responsible for over 95 percent of the road damage caused by traffic. In addition, weather damage represents a significant fraction of road maintenance in many states. Trucks should pay for a proportion of weather damage, through the appropriate ratio is debatable. In addition, highway user taxes made no fiscal contribution to the other necessary functions of society, unlike the property taxes paid by railroads on their roads, and no financial accounting is made of the roughly 5,000 people killed and thousands injured each year by trucks, or other costs, such as pollution, that burden society. The author reviewed the totality of expenses and the taxes paid by truckers, and believes that the subsidy to trucking is at least $30 to $40 billion per year, and perhaps higher.

The author further estimates that the overall savings from transferring 100 percent of heavy trucking to rail could approach $100 billion. Maintenance would be reduced to weather-related damage. The need for highway expansion would drop dramatically. Both air pollution and traffic injuries, including fatalities, would drop, reducing these social costs. Weight restrictions on bridges would have far less impact and would be more acceptable. The $61.6 billion in General Fund and other non-highway funding could be eliminated, and the road and highway infrastructure would still improve in relationship to the reduced demand.

6. According to John Schumann of LTK Engineering, who places the standard cost of uncomplicated single track at $2 million/mile, or $2.5 million/mile for double track. More complex installations, e.g., tunnels, bridges, mountainsides, etc., can cost $4–10 million/mile.

7. See: www.ble-t.org/pr/news/pf_headline.asp?id=25899.

8. This figure is based on Bureau of Transportation Statistics railroad and intercity trucking ton-mile data; see: www.bts.gov/publications/national_transportation_statistics/html.table_01_46a.html. According to BTS, railroads get 480 ton-miles per gallon of diesel (www.bts.gov/publications/national_transportation_statistics/html/table_04_17.html). Using this data, 80 percent of rail ton-miles plus 50 percent of trucking ton-miles would be equivalent to 4.279 billion gallons of diesel that is currently used to haul freight that could be transported on electrified tracks. Both diesel-electric and electric locomotives use electric motors as the final drive. However, the thermodynamic efficiency of the diesel engine-generators used to drive the electric motors on modern diesel-electric locomotives is in the high 30 percent range. By comparison, the grid-to-electric motor efficiency of electrified railroads is well over 90 percent (transmission, transforming, and rectifying losses); thus, there is a 2.5 ratio in energy efficiency between input diesel energy and input grid electrical energy—if the benefits of regenerative braking in electric locomotives are ignored. Converting the energy of 4.279 billion gallons of diesel to electricity, and dividing by 2.5, gives 69,660 GWh. In 2010, US energy generation was 4,125,060 GWh. So, by rough approximation, a massive switch of freight to electrified railroads as described in this chapter would require less than 1.7 percent of US electrical demand. However, this likely overstates the actual demand, since regenerative braking is not assumed and most of the truck freight would shift to double-stack container trains, further increasing efficiency.

9. This proposal was discussed at METRA Electrification and Commuter Rail Workshop, Chicago, March 2011. (Confirmed via personal communication with BNSF vice president Paul Nowicki, March 2011.)

10. S. Lothes, "Wired Up: The Future of Electrification on America's Freight Railroads," *Trains Magazine* (November 2009), www.railsolution.org/uploads/PDF/TRAINSarticle11-09.pdf.

11. This idea was discussed at METRA Electrification and Commuter Rail Workshop, Chicago, March 22, 2011 (see: www.rtachicago.com/meetings/electrification-and-commuter-rail-workshop.html).

12. J. Benes et al., "The Future of Oil: Geology versus Technology" (International Monetary Fund Working Paper, 2012), www.imf.org/external/pubs/ft/wp/2012/wp12109.pdf.

13. Personal correspondence with Ed Tennyson (during ongoing collaboration with author; see, for example: oilfreedc.blogspot.com/ as one example of our ongoing collaboration) and with Professor Emeritus Hans-Joachim Zierke (July 24, 2007).

14. Cost estimate from Drue Wands, rail supervisor of Boh Brothers Construction (personal communication, March 2007).

15. Association of American Railroads, "Positive Train Controls: Background Paper Estimates Total Costs at $13.2" (AAR report, 2011), www.aar.org/~/media/aar/Background-Papers/Positive-Train-Control-03–2011.ashx.

16. Rail fatality data retrieved from *Progressive Railroading* magazine: www.progressiverailroading.com/safety/article/Rail-fatalities-rose-in–2010-NTSB-data-shows—29254#. Trucking fatality data retrieved from TruckInfo, a trucking industry website: www.truckinfo.net/trucking/stats.htm#Accident%20stats.

17. Federal Railroad Administration, "Federal Track Safety Standards Fact Sheet" (FRA publication, US Department of Transportation, 2008), www.fra.dot.gov/downloads/PubAffairs/track_standards_fact_sheet_FINAL.pdf.

18. According to Gil Carmichael, former federal railroad administrator and currently senior board chairman for the Intermodal Transportation Institute at the University of Denver, quoted in R. Malone, "Railroads Can Move Forward," *Forbes*, May 5, 2006, www.forbes.com/2006/05/04/railroads-intermodal-shipping-cx_rm_0505rail.html?partner=alerts. According to Carmichael, a double-stack freight train can replace as many as 300 trucks and achieve nine times the fuel efficiency of highway movement of the same tonnage volume. The author's calculation of rolling and aerodynamic friction between trucks and double-stack container trains resulted in an energy ratio approaching 10 to 1 per container, while personal correspondence with Carmichael gave a ratio of 4 to 1 for trailer on flatcar (TOFC) versus trucking. The FRA gives a ratio of 3.2 to 1 for TOFC over trucking. Given the much lower payload to tare weight and higher aerodynamic resistance of trailer on flatcar, the author's calculations support a figure close to 4 to 1.

19. See, for example: T. Prince, "Towards an International Intermodal Network," *American Shipper*, November 2001, trid.trb.org/view.aspx?id=591513.

20. A bill of lading is a document issued by a carrier that lists goods being shipped and specifies the terms of their transport.

21. These figures are taken from American Association of Railroads press release, "Freight Railroads Expect to Spend a Record $13 Billion."

Part 3

Moving Forward

12 Imagining a Future Without Oil for Car-Dependent Cities and Regions

Peter Newman

The period from the 1930s to 2008 was the era of cities based on the Internal Combustion Engine (ICE). The Global Financial Crash (GFC) of 2008 and the issues of peak oil and climate change seem to have ended the domination of this technology, though it will take some time for it to phase out. The limitations of ICE technology have exercised the minds of many technologists and regulators who have struggled to make cleaner and greener cities that have less smog. But the biggest force driving the need to phase out ICE-based mobility is the problem of oil. As oil production reaches its decline phase, the overwhelming need to find more oil has led to more and more dangerous deep-sea oil wells and options like burning rocks filled with tar sands or deep fracking of trapped oil.

My reading of the trends in technology, global climate-change governance, peak oil, city planning, urban economics, and urban cultural change suggest that we have at this point in history a convergence toward a new kind of city building based around renewables and electric transport. It promises to create much cleaner and greener cities than could have been imagined before.[1] These cities will be oil-free. This chapter will outline how I see this unfolding.

The Historical Opportunity

At each point in industrial history, different waves of innovation have shaped our cities. As seen in figure 12.1, the waves of innovation set out by Hargroves and Smith can be seen to rise and then fall, with a major economic downturn punctuating each of the industrial phases.[2] My interest has been in how transport and associated fuels change

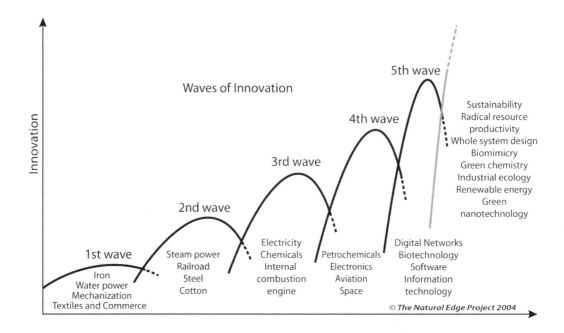

Figure 12.1 The waves of economic innovation showing the booms and busts related to how cities have adopted and built themselves around new technologies.

the nature of cities and hence facilitate these economic eras.[3] It can help us to see how the history of cities as well as the future of cities can be explained and how this can indicate the emergence of the next era of oil-free cities.

Early industrial innovation began in the old walking cities that were linked by water transport and began using water power to make industrial products. The limits to this began to be obvious in space and materials; thus, the next phase of innovation that arose from the global economic downturn of the 1840s saw the arrival of the steam engine and railway. Cities began to spread out along the rail tracks and to build much more production based on steam power and steel. By the 1890s the problems of steam as a power source—the air pollution of coal smoke and the ever-present danger of exploding boilers—made the advantages of newly harnessed off-site electric power obvious. This enabled electric trams and electric trains to spread the city along corridors and the production systems to be separate from their power source—and enabled, as well, the delights of brightly lit cities. The production of electricity multiplied the possibilities of coal-fired power enormously, but with growing consequences. By the 1930s these cities served by electric power were reaching their limits and a new era was created around the automobile, cheap oil, and highways that enabled cities to spread in every direction and much further out. And now these automobile-based cities are reaching their limits just as previous cities reached theirs. Cities can no longer afford the costs of their urban sprawl and the associated vulnerability of living off the oil that needs to be imported from highly unstable regions or highly dangerous environments. New kinds of cities are needed.

Where Are We Today?

Around the world there is a dramatic revival of electric public transport systems and the communities they serve. The Global Cities data currently being updated show that the historic decline in public transit use has reversed, and transit use is once again growing quite rapidly: in ten major US cities from 1995 to 2005 transit boardings grew 12 percent from 60 to 67 per capita; five Canadian cities grew 8 percent from 140 to 151; four Australian capital cities rose 6 percent from 90 to 96 boardings per capita; and four major European cities grew from 380 to 447 boardings per capita, or 18 percent.[4]

Fast rail between cities is going through a rapid growth phase even as aviation struggles with oil prices—especially in China and Europe. Japan built their system in response to the early oil crises, and now other countries like the United States and Australia are recognizing that they must follow suit, even though this process of transition is painfully slow in an era when major infrastructure investment is difficult.[5]

Car-based city building has created shopping malls and dormitory suburbs rather than the interactive, walkable city centers that are needed for the service-oriented knowledge economy. As traditional city centers have recovered and grown with these new service jobs, they have also attracted residential development for those wanting a more urban lifestyle. Now the need for these centers has shifted to the middle and outer suburbs, and the basis of this shift seems to be good rail systems that can attract development around them—transit-oriented development (TOD). As outlined below, transit-oriented development must also be Pedestrian-Oriented Development (POD) and Green-Oriented Development (GOD) in order to make the fullest contribution to creating the new oil-free economy in our cities and regions.

Data on the growth in European, Australian, and American city transit systems (and the decline in car use) is set out in fig. 12.2.[6] More-recent data on Australian cities confirms that "peak car use" has been reached, and those cities which have invested in their public transport systems (especially electric rail) in recent decades (Melbourne, Brisbane, and Perth) have shown substantial growth in transit. Cities without electric rail (like Canberra, Hobart, and Darwin) have all experienced declines in car use (despite having virtually no congestion issues) but also declines in transit use.

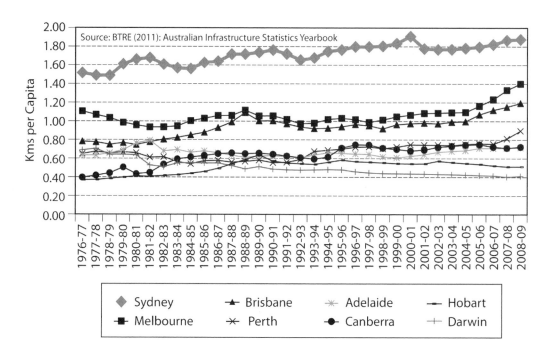

Figure 12.2   Recent trends in public transport in Australian cities. (Source: Glazebrook, 2011.)

So What Comes Next?

The limit of oil, with its associated need to dig deeper and more dangerous deposits, the limit of fossil fuels in general due to their contribution to climate change, and the limits facing cities as they continue to sprawl and remain automobile dependent, is now apparent. Climate-change governance at every level is now requiring plans be implemented to reduce greenhouse gases; the global goal for this reduction is 80 percent by 2050. This has undermined most long-term planning based on the fourth-wave economy, in which fossil-fuel use expanded exponentially. Peak oil happened in 2008 (at least for conventional oil sources)[7] when oil prices tripled and the areas where cities had spread or indeed scattered beyond normal commuting times simply crashed as they had few local services, few local jobs, and no public transport. Areas with the greatest car dependence were hardest hit by the sub-prime-mortgage melt-down. Many of these suburbs are now largely abandoned and some cities are even contemplating bulldozing urban areas that are beyond any conceivable repair.

Added to the slow economic recovery for auto-dependent areas are the signs in urban areas of decreasing car use, increasing transit options, increasing density, and more-limited urban highway building (all of which are discussed below), demonstrating the beginning of a market-based change that is setting into car-dependent cities.[8] It seems like one era is ending and another is dawning.

The next phase in the innovation waves, according to Hargroves and Smith in figure 12.1, is a combination of the digital economy and the sustainability economy. The sustainability innovations all require the benefits of information and communications technology (ICT). In terms of transport, it seems that the main innovations are electric transit and electric vehicles based on renewable energy.[9] What are the options and what will they mean for our cities and regions?

In combination with technologies including electric transit and electric vehicles, our cities will be shaped by such practices as travel demand reduction, travel demand management, and investment in regional biofuels and renewable natural gas.

1.   Investing in Rail Transit

This section will show how electric transit powered by renewable energy can bring about dramatic reductions in car use and the associated need for oil. This is a correction to the total automobile dominance in the past decades of planning and city building because it is socially and economically as well as environmentally limited.

When cities provide a combination of transportation and land-use options that are favorable for green modes, and offer time savings compared to those of car travel, then the switch to transit is inevitable. This means that wherever transit is reasonably competitive with car traffic in major corridors, then people will use it, especially as fuel prices continue their inexorable rise at around 5 percent per year. Those cities where transit is relatively fast are those with a reasonable level of support for it.[10] The reason is simple—they can save time. Perth was the first Australian city to begin this transition to modern fast rail,[11] and its successful model has now become the basis for rail growth in other Australian cities, especially given the new Federal Government funds from Infrastructure Australia, 56 percent of which are slated for urban rail.[12]

With fast-rail systems, the best European and Asian cities—that is, those with the highest ratio of transit to traffic speeds—have achieved a transit option that is faster than the car down the main city corridor. Rail systems are faster in every city in our 84-city sample by 10–20 kilometers per hour (kph) over bus systems, as buses rarely average more than 20–25 kph.[13] Busways with a designated lane can be quicker than traffic in car-saturated cities, but in lower-density car-dependent cities it is important to use the extra speed of rail to establish an advantage over cars in traffic. This is one of the key reasons why railways are being built in over a hundred US cities that shut down high-quality rail in the beginning of the Fourth Wave and are now regretting their unbalanced transport systems.

Rail has a density-inducing effect around stations, which can help to provide the focused centers so critical to overcoming car dependence. Thus, transformative change of the kind that is needed to rebuild car-dependent cities comes from new electric-rail systems as they provide a faster option than cars and can help build transit-oriented centers.

We need to stop increasing road capacity and instead provide major increases in transit capacity if we are going to help this transition to a clean, green, oil-free city and its associated economy. This won't be easy, given the funding systems that have developed around the car and truck. However, evidence from the United States shows that car use began to spiral downwards starting in 2004 (as it did in Australian and European cities) and had its biggest drop in 50 years in 2009 at 4.3 percent per year; meanwhile, transit use increased 6.5 percent, also a record.[14] The transition process seems to be underway, with urban freeway building almost stopped in US cities.

How much is it possible to change our cities? Is it possible to imagine an exponential decline in car use in our cities that could lead to a 50 percent decrease in passenger-kms driven in cars? The key mechanism for change is a quantitative leap in the quality of public transport while fuel prices continue to climb, accompanied by an associated change in land-use patterns.

Figure 12.3 shows the relationship between car passenger-kms and public transport passenger-kms from the Global Cities Database.[15] The most important thing about this relationship is that as the use of public transport increases linearly the car passenger-kms decrease exponentially. This is due to a phenomenon called *transit leverage* whereby one passenger-km of transit use replaces between three and seven passenger-kms in a car due to more-direct travel (especially in trains), trip chaining (doing various other things like shopping or service visits associated with a commute), giving up one car in a household (a common occurrence that reduces many solo trips), and eventual changes in where people live as they prefer to live or work nearer transit.[16]

The data on private transport use and public transport use in selected Australian cities for 1996 is given in table 12.1 (passenger kilometers per capita in each case). The values in figure 12.3 compared to those in table 12.1 show that Australian cities are somewhat down the curve from the very high values of those US cities that have almost no transit (some around 100–200 passenger-kms per person) and very high private transport use of over 15,000 passenger-kms per person. The data show that Sydney is the Australian city with the highest transit use, with 12.6 percent of its total motorized passenger-kms traveled on transit, and that the lowest was Perth, with 4.5 percent (this was before the remarkable increase in patronage associated with Perth's rail revival).

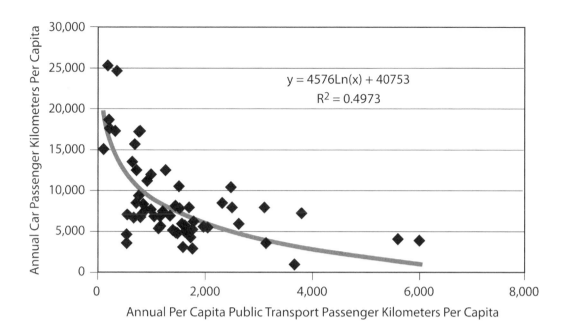

Figure 12.3   Car travel vs. public transport use in global cities, 1995. (Source: J. Kenworthy and F. Laube, *The Millennium Cities Database for Sustainable Transport* [UITP publication, Brussels, 2001].)

Table 12.1 Car and public transport use per capita in four Australian cities, 1996

City	Private transport use (passenger-kms/person)	Public transport use (passenger-kms/person)	Percentage of public transport as a share of total transport
Sydney	10506	1509	12.6%
Melbourne	11918	994	7.7%
Brisbane	12487	720	5.5%
Perth	13546	642	4.5%

Source: J. Kenworthy and F. Laube, *The Millennium Cities Database for Sustainable Transport* (UITP publication, Brussels, 2001).

If Sydney doubled its transit use to 3,018 passenger-kms per person, it would have a per-capita private-transport use of 4,088 passenger-kms per capita—a 61 percent reduction in car passenger-kms per person over the 1996 figure. If Perth is able to continue the rapid growth in transit patronage and triple its 1996 use to around 2,000 passenger-kms per person, then it will reduce its private-transport use to 6,000 car passenger-kms per capita, which is a reduction of 56 percent over the 1996 level. Similar calculations can be done for the other Australian cities. Indeed, it is feasible that each city could set a target of increases in passenger-kms per capita for public transport in order to achieve certain target reductions in car use as part of their commitment to reaching the national goal of 80 percent reduction in greenhouse gases by 2050.

The biggest challenge in an age of radical resource-efficiency requirements will be finding a way to build fast-rail systems back into scattered, car-dependent suburban areas. Rail reserves are not available, so many cities are choosing to tunnel or go above ground on elevated structures; both are expensive. The solution may well be provided by Perth and Portland, Oregon, which have built fast-rail systems down freeways, deep into car-dependent suburbs that were really hurting in the recent fuel price hike. To build fast electric rail down the middle of roads is easier than anywhere else, as the right-of-way is there and engineering in terms of gradients and bridges is compatible. If a median is not available, then a lane may need to be replaced by rail, but in an era of declining car use this should be possible. Such expedients are not ideal in terms of building transit-oriented developments (TOD), but they can still be employed by using high-rise buildings as sound walls. Linkages from buses, electric bikes, and park-and-ride are all easily provided so that local travel to the system is short and convenient.

The speed of the transit system is key, and in Perth the new Southern Railway has a maximum speed of 130 kph (80 mph) and an average speed of 90 kph (55 mph)—that is, at least 30 percent faster than auto traffic. The result is dramatic increases in patronage, far beyond the expectations of planners who see such suburbs as too low in density to deserve a rail system. The Southern Suburbs Rail line opened in December 2007 and is now carrying 65,000 people a day, whereas the bus system carried just 14,000. There is little else that can compete with this kind of option for creating a future in the car-dependent suburbs of many cities.

Fast electric-rail services are not cheap. However, they cost about the same per mile as most freeways, and for the past fifty years we have been able to find massive funding sources for building freeways. In the transition period it will require some creativity, as the systems for funding rail are not as straightforward. In Perth, the state government was able to find all the funds from the treasury due to a mining boom and was even able to pay off the entire rail system, including the new Southern Railway even before it was opened. But for most cities this is not possible. Infrastructure Australia has stepped into the breach and provided $4.6 billion in federal government funding for urban rail systems—mostly in Melbourne, Adelaide, and Gold Coast. Planning studies indicate that Sydney and Brisbane will also build in a doubling of rail capacity. This is an historic commitment as federal funds in recent decades have only gone to roads.[17]

The funding solution will also require cities to find innovative partnerships, such as financing transit through the use of taxes or direct payments from land development as in Copenhagen's new rail system, or through a congestion tax as in London. Funding of transit in congested cities can occur as it has in Hong Kong and Tokyo, where the intensive development around stations means that the transit can be funded almost entirely from land-redevelopment value capture. In poorer cities, the use of development funds for mass transit can increasingly be justified through the transformation of their urban economy. Peak oil and climate change will increasingly be part of that rationale.[18]

2. *Reducing Travel Miles*

There are many ways that technology can help us to reduce the need to travel. Long-distance travel or even short-distance trips within cities may be considerably reduced once the use of broadband-based telepresence begins to make high-quality imaging feasible on a large scale. There will always be a need to meet face-to-face in creative meetings in cities, but for many routine meetings the role of computer-based meetings will rapidly take off. Cities that are attractive places to meet (walkable, safe, and lively) will thrive even more in these conditions, while suburban areas with little more than scattered houses will find their economies being undermined.

The need to increase densities in car-dependent cities has been recognized for many decades, but recent "peak car" data suggest that densities are increasing as younger people and empty-nesters come back into urban areas.[19] The return to cities reverses the declines in density that have characterized the growth phase of automobile cities in the past 50 years. Table 12.2 contains data on a sample of cities in Australia, the United States, Canada, and Europe, showing urban densities from 1960 to 2005 that clearly demonstrate this turning point in the more highly automobile-dependent cities. In the small sample of European cities, densities are still declining due to *shrinkage*, or absolute reductions in population, but the data clearly show the rate of decline in urban density slowing down and almost stabilizing as re-urbanization occurs.

Table 12.2 Trends in urban density in some US, Canadian, Australian, and European cities, 1960–2005

Cities	1960 Urban density persons/ha	1970 Urban density persons/ha	1980 Urban density persons/ha	1990 Urban density persons/ha	1995 Urban density persons/ha	2005 Urban density persons/ha
Brisbane	21.0	11.3	10.2	9.8	9.6	9.7
Melbourne	20.3	18.1	16.4	14.9	13.7	15.6
Perth	15.6	12.2	10.8	10.6	10.9	11.3
Sydney	21.3	19.2	17.6	16.8	18.9	19.5
Chicago	24.0	20.3	17.5	16.6	16.8	16.9
Denver	18.6	13.8	11.9	12.8	15.1	14.7
Houston	10.2	12.0	8.9	9.5	8.8	9.6
Los Angeles	22.3	25.0	24.4	23.9	24.1	27.6
New York	22.5	22.6	19.8	19.2	18.0	19.2
Phoenix	8.6	8.6	8.5	10.5	10.4	10.9
San Diego	11.7	12.1	10.8	13.1	14.5	14.6
San Francisco	16.5	16.9	15.5	16.0	20.5	19.8
Vancouver	24.9	21.6	18.4	20.8	21.6	25.2
Frankfurt	87.2	74.6	54.0	47.6	47.6	45.9
Hamburg	68.3	57.5	41.7	39.8	38.4	38.0
Munich	56.6	68.2	56.9	53.6	55.7	55.0
Zurich	60.0	58.3	53.7	47.1	44.3	43.0

The relationship between density and car use is also exponential, as shown above. If a city begins to slowly increase its density, then the impact can be more extensive on car use than expected. Density is a multiplier on the use of transit and walking/cycling as well as reducing the length of travel. Increases in density can result in greater mixing of land uses to meet peoples' needs nearby. This is seen, for example, in the return of small supermarkets to the central business districts of cities as residential populations increase and demand local shopping opportunities within an easy walk. Overall, this reversal of urban sprawl will undermine the growth in car usage.

The need to develop city centers is now shifting to the need to develop sub-centers across the whole car-dependent city in order to help get people out of cars. This process aims to create transit-oriented development (TODs). This will ultimately lead to the polycentric city—a series of sub-centers or small cities in the suburbs, all linked by quality transit but each providing the local facilities of an urban center.

The facilitation of TODs has been recognized by all Australian cities and most American cities in their metropolitan strategies.[20] The major need for TODs is not in the inner areas, as these have many TODs from previous eras of transit building, but in the newer outlying suburbs. There are real equity issues here, as the poor increasingly are trapped on the fringe, with high expenditures on transport. A 2008 study by the Center for Transit Oriented Development shows that people in TODs drive 50 percent less than those in conventional suburbs.[21] In both Australia and the United States, homes that are located in TODs are best at holding their value or appreciating the fastest under the pressure of rising fuel prices. The report suggests that TODs would appreciate fastest in up-markets and hold value better in down markets. This is the rationale for how TODs can be built as public-private partnerships (PPPs) in rail projects.[22]

Thus, TODs are an essential policy for responding to peak oil, especially when they incorporate affordable housing. The economics of this approach have been assessed by the Center for Transit Oriented Development and the NGO Reconnecting America. In a detailed survey across several states, these NGOs assessed that the market for people wanting to live within half a mile of a TOD was 14.6 million households. This is more than double the number who currently live in TODs. The market is based on the fact that those now living in TODs (who were found to be the same age and the same income on average as those not in TODs, and living in smaller households than those not in TODs) save some 20 percent of their household income by not having to own so many cars; indeed, those in TODs owned 0.9 cars per household compared to 1.6 outside. This frees up an average of $4,000–5,000 per year. In Australia, a similar calculation showed this would save some $750,000 in superannuation over a lifetime. Most importantly, this extra income is spent locally on urban services, which means the TOD approach is a local economic development mechanism.[23]

TODs must also be PODs—that is, pedestrian-oriented development—or they lose their key quality as a car-free environment to which businesses and households are attracted. Urban designers need to assure that public space is vibrant, safe, and inviting. Jan Gehl's transformations of central areas such as Copenhagen, New York, London, and Melbourne are showing the principles of how to improve TOD spaces so that they are more walkable, economically viable, socially attractive, and environmentally significant.[24] Gehl's work in Melbourne and Perth has been evaluated after a decade of implementation and in both cases it indicates substantial increases in walkability, numbers of pedestrians who use the city center for their work, shopping, education, and especially recreation.[25] It will be important for those green developers wanting to claim credibility that scattered urban developments, no matter how green in their buildings and renewable infrastructure, will be seen as failures in a post peak-oil world unless they are pedestrian-friendly TODs.

At the same time, TODs that have been well designed as PODs will also need to be GODs—green-oriented developments. TODs will need to ensure that they have full solar orientation, are renewably powered with smart grids, have water sensitive design, use recycled and low-impact materials, and use innovations like green roofs.

Examples of TOD-POD-GODs are appearing in many places but an early good example is the redevelopment of Kogarah Town Square in Sydney. This inner-city development is built upon a large city council car park adjacent to the main train station, where there was once a collection of poorly performing businesses adjacent. The site is now a thriving mixed-use development consisting of 194 residences, 50,000 square feet of office and retail space, and 35,000 square feet of community space including a public library and town square. The buildings are oriented for maximum use of the sun with solar shelves on each window (enabling shade in summer and deeper penetration of light into each room), photovoltaic (PV) collectors are on the roofs, all rain water is collected in an underground tank to be reused in toilet flushing and irrigation of the gardens, recycled and low-impact materials were used in construction, and all residents, workers, and visitors to the site have a short walk to the train station (hence reduced parking requirements have enabled better and more productive use of the site). Compared to a conventional development, the Kogarah Town Square saves 42 percent of the water and 385 tons of GHG—this does not include transport-oil savings that are hard to estimate but are likely to be even more substantial.[26]

While the demand for TODs is growing, creating TODs can still present significant challenges, given both the complexity of financing TODs and also the number of private and public actors involved. TODs are in great demand, which often results in housing being priced out of the range of middle- and lower-income households. Thus, along with the other green requirements for TODs there needs to be a requirement for a certain proportion of affordable housing. In Perth, it has been suggested that the 20 or so TODs being planned be progressed via a new TOD zoning that requires minimal amounts of parking, maximizes density and mix, includes green innovations, and has a minimum of 15 percent affordable housing to be purchased by social housing providers.

3. Reducing Demand for the Automobile

The real test of a 100 percent oil-free city will be how it can simultaneously reduce its oil consumption through new technologies and reduce travel demand through urban design, while at the same time changing behavior in order to reduce the demand for car travel. The need for a parallel delivery of physical infrastructure and "mental infrastructure" will be essential for any city contemplating the development of an oil-free city.

Every city and every nation has its own way of making the adoption of more planetary lifestyles convenient and easy compared with more consumptive lifestyles. When it comes to cars, however, the more that a city is car-dependent, the harder it is to use tax incentives to change people's lifestyles. European cities have much higher gasoline taxes than American and Australian cities, and accordingly they have less car usage.[27]

In the car-dominated cities of North America and Australia, the major public policy to reduce the global and local impacts of the car has been regulations requiring cleaner vehicles. Following introduction of such requirements, most urban atmospheres have become cleaner, although fuel use has continued to increase as vehicles become bigger and their use continues to grow. Regulations are also applied to safety and congestion management, but congestion will continue to worsen if more and more car use is facilitated. Regulations alone do not change behavior. The economic principle known as the Jevons Paradox—increasing efficiency means increasing consumption—has been found to apply to car use. If people buy cars that use less fuel, they just drive them more—undermining most gains made possible through new technology. Prices do affect behavior and the rise in fuel prices is taking the world into new territory, though elasticities tend to be quite small.[28]

All these necessary policy approaches will be wasted without education on a changed role for the car. Thus, cultural change to help people reduce their desire to drive needs to be part of any city's policy arsenal if it is to face up to the challenge of growing a sustainable city. TravelSmart is one such program that shows this is indeed possible.[29]

German sociologist Werner Bróg developed TravelSmart based on the belief that behavior change toward less car dependence can work—if it is community based and household oriented. After some trials in Europe, TravelSmart was adopted in large-scale projects in Perth, Western Australia. It has since spread across most Australian cities and to other European cities, especially in the United Kingdom, and has now been piloted in six American cities.

Good behavior-change programs do not use media. Instead, TravelSmart targets individual households directly, asking them if they would like to find ways to reduce their car use. The residents who show interest receive information and, for the few who need extra support, a visit from a TravelSmart officer or eco-coach who encourages people to start with local trips, especially the school trip for children. Walking to school is now seen as an essential part of the healthy development of young people's sense of place and belonging in any community as well as a way to reduce obesity.[30]

TravelSmart has been found to reduce the vehicle-kilometers traveled by around 12–14 percent across whole communities—a result that seems to last for at least five years after the program ends. Where public transit is not good and destinations are more spread out, the program may only reduce car use 8 percent, but where these are good it can rise to 15 percent.[31]

Behavior-change programs can develop social capital to support people as they try to get out of their cars, and this can change local cultures. People show friends how much money it saves as well as making them feel they are doing their bit to reduce climate change and oil vulnerability. There is evidence in Brisbane, Australia, that at least 50 percent more people than those involved in the initial household interviews were actually following the program when the surveys were first done; in other words, people were spreading the message to their friends and colleagues.[32] When people start to change their lifestyles and can see the benefits, they become advocates of sustainable transport policies in general. The politics of change is easier to manage when communities themselves have begun to change.

In Perth, the extension of the rail system to far outer suburbs has been more positive and politically achievable than expected, with a massive 90 percent support for the last stage, the Southern Suburbs Railway. In parallel to this political process, Perth had some 200,000 households undergoing the TravelSmart program; the suburbs where TravelSmart had been conducted increased their public-transport patronage by 83 percent, while in areas without TravelSmart it was just 59 percent.[33]

TravelSmart has now spread across the world and some of the data from cities in Europe and the United States confirm the kind of changes that can be expected from this grassroots approach (fig. 12.4).

TravelSmart has now reached more than 450,000 residents in Perth, at a cost to the state of under A$36 per resident. Worldwide, TravelSmart's individualized approach to travel-demand management has been delivered to approximately five million people. If you take into account the reductions in the public and private costs of car use that it has achieved, the program saves $30 for every dollar it costs.[34]

Behavior-change programs can also work in the workplace. TravelSmart was found to work well when a TS Club was formed that enabled people to share experiences, bring in local speakers, and lobby for facilities like showers for bike riders and transit passes instead of parking spaces. For example, the natural gas company Woodside in Perth involved their employees in planning their new building; a strong representation from the TS Club at Woodside led to good bicycle facilities being provided. The

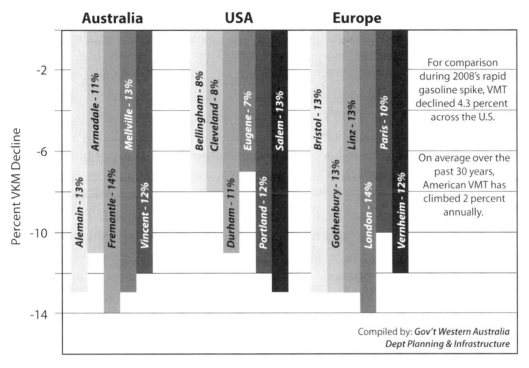

Figure 12.4   Reductions in car vehicle-kilometers (VKM) traveled in cities with TravelSmart programs.

firm now has more employees biking to work than driving, and the subsequent saving in car parking spaces is considerable.[35]

TravelSmart has also used the Walking School Bus concept to enable children to get to school safely. A variation is the bicycle bus in which adults supervise children riding their bikes to school. Walking or cycling to and from school provides valuable exercise for children. It enables them to learn early on that walking and cycling are very practical, pleasant, and healthy forms of local travel. At the same time, the adult supervision that Walking School Bus schemes provide ensures that children—especially very young ones—are safe, thus overcoming a barrier that causes parents to prohibit their children from walking or cycling. The experience of walking or cycling teaches children about their neighborhood and environment, as well as giving them road-safety skills and equipping them for independent mobility as they get older. Finally, these programs, if they are supervised by parents or volunteers, involve virtually no costs.[36]

The changes in behavior being seen among young people in cities across the world, especially in the United States, suggest that programs to assist this process may have considerable success in reducing the demand for car use.[37] Much of this cultural shift appears to be because young people are connected substantially through social media devices rather than cars, and to a growing use of Skype and telepresence to replace the need for travel.[38]

4. Plugging-In to Electric Vehicles

Even if, by a rather herculean effort, we manage to reduce car use by 50 percent as suggested above, we will still have to reduce the oil consumed and carbon emitted by the other 50 percent of vehicles being used. The question should therefore be asked: What is the next-best transport technology for motor vehicles? The growing consensus seems to be: plug-in electric vehicles (PEV). Plug-in electric vehicles are now viable alternatives due to new batteries such as Lithium Ion and hybrid engines for extra flexibility. PEVs are likely to be attractive to the market in the transition period as electric-recharging infrastructure builds up. For clean, green cities, plug-in electric vehicles not only reduce oil vulnerability but are also becoming a critical component in the energy-renewability of a city's electricity grid. PEVs can serve a storage function for renewables through vehicle-to-grid linkages within a smart grid. Thus, electric vehicles are becoming an essential part of the effort to make cities clean and green and oil-free.[39]

When electric vehicles are plugged in to be recharged they can be a part of the peak-power provision. Peak power is the expensive part of an electricity system, and suddenly renewables are offering the best and most reliable option as well as increasingly the cheapest option. This technology, electric vehicle batteries being linked in to the electricity grid, is known as vehicle-to grid (V2G) technology and requires the digital systems controls of a smart grid. By enabling a much higher proportion of a city to be renewably powered, the option of electric vehicles—electric buses, electric bikes, scooters and gophers, and electric cars—will have an important role in the future oil-free city.[40]

Electric rail can also be powered from the sun, either through the grid powering the overhead wires or in the form of new battery-based light rail (without overhead catenaries) that could be built along highways into new suburbs. The first example of this technology is now running in Bordeaux, and the next-generation light rail is likely to be battery-based with electric power through high-powered contactless charging at stations. These will work better if the stations are green developments with renewable power built into their fabric and available for quick recharging services to trains and to PEVs, which can also help in local power storage.

Signs that this transition to electric transport is under way are appearing in demonstration projects such as those in Boulder and Austin, in Google's 1.6 MW solar campus in California (with 100 PEVs), and in the fact that oil companies are acquiring electric utilities.

What sort of immediate impact could there be? According to one study, the integration of hybrid EV cars with the electric power grid could reduce gasoline consumption by 85 billion gallons per year. That's equal to:

- 27 percent reduction in total US greenhouse gases,
- 52 percent reduction in oil imports, and
- $270 billion not spent on gasoline.[41]

Al Gore has called the smart grid/renewables/EV transition the "moonshot" of our era, as it has the potential to enable 100 percent renewables in a decade.[42]

5. *Investing in Renewable Fuels*

Biofuels have promised a lot but have been criticized for their impact on food prices when grain is converted to fuel. However, they still have a potentially significant role in some areas where there is a crop surplus, and also, eventually, when the technology improves to make them from cellulose materials (agricultural and forestry waste) and from blue-green algae. It is likely that biofuels will be used as a do-it-yourself fuel on farms.[43] Thus, biofuels may have a role in agricultural regions as a fuel to assist farmers, but as a widespread fuel for cities they do not represent an option that can be yet taken seriously.

Aviation will not easily cope with the rapid rise in fuel price from the peak-oil/carbon-pricing double whammy. At the height of the 2008 fuel crisis there was panic among airlines as the price of fuel rose to more than 50 percent of the price of a ticket.[44] Gilbert and Perl do suggest a few ways that air travel will adapt, but mostly they see little of potential other than regional high-speed rail and a return to ship travel.[45] Others such as the International Energy Agency (IEA) see biofuels as the only real option for aviation.[46] But the cost of biofuels for aviation appears to make them prohibitive at this stage.

Perhaps the regional transport technology that could make a comeback is airships. These are able to fly at low levels at speeds of 150–200 kph and carry large loads while using just one-tenth of the fuel of fixed-wing aircraft. Airships are already being used to carry large loads to remote mining areas and to take groups of about 200 people on eco-tourism ventures, just as a cruise ship might.[47] Perhaps aviation is a possible major use for biofuels, as no other obvious oil-free option is presenting itself for this part of the transport task.

What do you do with freight transport and regional transport outside of cities where electric grids are not so easily used with vehicles? And what about industries that presently use oil? Can they also go oil-free?

There will almost certainly be a reduction in the amount of freight moving around as fuel prices eat into the transport economics of consumption (see chap. 11). Containers will be reduced as their fuel costs move from being 10–15 percent to over 50 percent of their total transport costs. Food-miles will start to mean something to food prices when the cost of fuel triples. But trucks and trains and regional transport will still go on.

There are various futures being predicted for freight, including the use of electric trucks powered through overhead catenaries like trams, though this is unlikely given the extent of the infrastructure required. In my view, the transitional stage for larger vehicles as well as for industry and regional transport will involve a switch to the greater use of natural gas and perhaps some biofuels. Trucks and trains and fishing boats can use CNG (compressed natural gas) or LNG (liquefied natural gas) in their diesel engines (with pay-off times of just a few years, due to high diesel costs). Cars for regional transport can also be switched over to natural gas as well (particularly if the manufacturer makes them standard, as occurred in Sweden when the government committed to natural gas cars for their vehicle fleet). The attraction is that natural gas infrastructure is already in place, with almost 80 percent of the population in most developed countries having access to reticulated natural gas.

Global natural gas production has had a boost in recent years due to the technology of fracturing shale, though not without the attendant environmental controversies. Thus estimates on its peak have been extended out into a long-distance future. However, the cost will be increasing with these new sources as well as with the use of more and more offshore gas, just as with offshore oil. But ultimately the issue of carbon reduction will undermine the long-term future of fossil natural gas. In reality, natural gas can only be a small part of the transitional arrangements for oil; it cannot be seen as the long-term replacement. It will, however, be an obvious way to ease the pressure on diesel supplies, and this will be a great advantage to ensure cleaner air in cities (gas buses have already shown their big advantages over diesel buses in air quality) as well as energy security and taking pressure off the need for oil production in dangerous places.

The benefit of the transition to natural gas has always been seen as an enabler of the long-term transition to hydrogen as a fuel source. However, there is another process that has much greater potential than the widespread use of hydrogen: the use of renewable natural gas. Biogas can be created from biomass via gasification and this could play a role in the future. However, there is an even bigger process that could assist in creating an oil-free future: the catalytic process known as the Satanalia Process facilitates the joining of CO_2 and H_2O into CH_4 and O_2, just as in the photosynthetic process whereby sunlight uses chlorophyll to catalyze CO_2 and H_2O into carbohydrates. Around the world a range of research labs are vying to develop catalysts that can turn this process into a commercially successful system.

Thus, if there can be a development of the hydrogenation of CO_2 using renewable energy, then natural gas can become a renewable fuel in itself and can be fed into the present natural gas grids and can even be an export product through LNG technology.[48] There is large potential in this process, considerably more than "clean coal," as a totally new infrastructure will not be required for its distribution (most coal fields and hence coal-fired power are long distances away from the deep caverns that can act to absorb CO_2). In the interim years, as coal-fired power stations are being phased out, they can have renewable natural gas production facilities attached to them. Eventually CO_2 can be extracted from the atmosphere and used as a renewable fuel.

Thus natural gas can be given a long-term future and can be part of the "oil-free" transition. Freight, industry, and regional transport are likely to continue to expand into natural gas and to transition into the use of renewable natural gas.

Conclusions

There are few guidelines to the future of our cities and regions that take account of what could happen to transport in response to the triple challenge of air pollution, climate change, and peak oil, or especially the growing awareness of the dangers associated with desperately searching for more oil. First, this oil-free future must be imagined based on what we know about technology, urban planning, and behavior-change possibilities available or emerging now. If we do not go through this process of imagination we can get submerged under an avalanche of despair as people can see only disaster once the days of cheap oil are over—and they are. It is therefore understandable why some people get so upset about the possibilities of collapse as suggested by Jared Diamond.[49] As Lankshear and Cameron say:

> Peak oil has already become a magnet for post-apocalyptic survivalists who are convinced that Western society is on the brink of collapse, and have stocked up tinned food and ammunition for that coming day.[50]

The alternatives all require substantial commitment to change in both how we live and how we use technologies in our cities and regions. The time to begin the changes we need is now, as they will take decades to get in place, and the time to respond to peak oil and climate change is of the same order, probably less for oil. But at least by imagining some of these changes, as suggested above, it is possible to see how we can get started on the road to oil-free cities.

The first signs of change toward these emerging technologies can now be seen: the dramatic growth in electric transit; the rapid move toward electric vehicles and smart grids with a 40 percent per annum growth in global renewables; the emerging use of natural gas and biofuels; and new technologies like Skype and telepresence. Their application into large-scale urban demonstrations is now under way in places like Kronsberg and Vauban in Germany, Masdar in the United Arab Emirates, and the low-carbon cities of China.

The potential for creating oil-free cities is here. The technologies and practices outlined above suggest that we can be oil-free by 2050 and renewables-based oil-free by 2100. We first need to imagine the changes that are available now in transport and urban design, and then begin the process of change through large-scale demonstrations. I remain hopeful that the oil-free city is not only possible to imagine but is a viable and attractive option.

Notes

1. P. Newman, T. Beatley, and H. Boyer, *Resilient Cities: Responding to Peak Oil and Climate Change* (Washington, DC: Island Press, 2009).

2. C. Hargroves and M. Smith, *The Natural Advantage of Nations* (London: Earthscan, 2004).

3. P. Newman and J. Kenworthy, *Sustainability and Cities: Overcoming Automobile Dependence* (Washington, DC: Island Press, 1999).

4. J. Kenworthy and F. Laube, *The Millennium Cities Database for Sustainable Transport* (UITP publication, Brussels, 2001); P. Newman and J. Kenworthy "Peak Car Use: Understanding the Demise of Automobile Dependence," *World Transport Policy and Practice* 17, no. 2 (2011): 32–42.

5. R. Gilbert and A. Perl, *Transport Revolutions: Making the Movement of People and Freight Work for the 21st Century* (London: Earthscan, 2006).

6. Bureau of Infrastructure, *Traffic Growth: Modelling a Global Phenomenon* (Report 128, [Australian] Transport and Regional Economics [BITRE], Canberra ACT, 2012).

7. D. Murray and D. King, "Climate Policy: Oil's Tipping Point Has Passed," *Nature* 481 (2012): 433–5.

8. Newman and Kenworthy, "Peak Car Use," 32–42.

9. See: P. Newman and R. Wills, "King Coal Dethroned," *The Conversation*, May 14, 2012.

10. J. Kenworthy, F. Laube, P. Newman, P. Barter, T. Raad, C. Poboon, and B. Guia, *An*

International Sourcebook of Automobile Dependence in Cities, 1960–1990 (Boulder, CO: University Press of Colorado, 1999).

11. The new Southern Rail grew in patronage by 19 percent in 2011 and reached its predicted numbers for 2026. Source: Department of Transport, Western Australia.

12. Commonwealth of Australia (2009), *Nation Building for the Future*, Budget Papers (report of the Department of Treasury and Finance, Canberra, May 2009).

13. Newman and Kenworthy, *Sustainability and Cities*.

14. R. Puentes and A. Tomer, *The Road Less Traveled: An Analysis of Vehicle Miles Traveled Trends in the US* (Brookings Institution publication, Metropolitan Infrastructure Initiative Series, Washington, DC, 2009).

15. Kenworthy and Laube, *The Millennium Cities Database*.

16. Newman and Kenworthy, *Sustainability and Cities*.

17. P. Laird, P. Newman, and M. Bachels, *Back on Track: Rethinking Transport Policy in Australia and New Zealand* (Sydney: UNSW Press, 2001).

18. R. Cervero, "Transit-Oriented Development in America: Strategies, Issues, Policy Directions," *New Urbanism and Beyond: Designing Cities for the Future*, ed. T. Haas (New York: Rizzoli, 2008), 124–9.

19. P. Newman and J. Kenworthy, *Cities and Automobile Dependence* (Aldershot, UK: Gower, 1989); Newman and Kenworthy, "Peak Car Use," 32–42.

20. C. Curtis, J. Renne, and L. Bertolini, *Transit Oriented Development: Making It Happen* (London: Ashgate, 2009).

21. Center for Transit Oriented Development and Reconnecting America, *Hidden in Plain Sight: Capturing the Demand for Housing Near Transit* (CTODRA publication, 2004), www.reconnectingamerica.org.

22. J. Mcintosh, P. Newman, T. Crane, and M. Mouritz, *Alternative Funding Mechanisms for Public Transport in Perth: The Potential Role of Value Capture* (publication of the Committee for Perth [Australia], 2011); Blake Dawson, "The New World of Value Transfer PPPs," *Infrastructure: Policy, Finance and Investment*, May 2008, 12–13.

23. H. Dittmar and G. Ohland, eds., *The New Transit Town* (Washington, DC: Island Press, 2004); CTODRA, *Hidden in Plain Sight*.

24. J. Gehl, *Life Between Buildings: Using Public Space*, trans. Jo Koch (New York: Van Nostrand Reinhold, 1987); J. Gehl and L. Gemzøe, *New City Spaces* (Copenhagen: Danish Architectural Press, 2000); J. Gehl, L. Gemzøe, S. Kirknaes, and B. S. Sondergaard, *New City Life* (Copenhagen: Danish Architectural Press, 2006); J. Gehl, *Cities for People* (Washington, DC: Island Press, 2010).

25. J. Gehl et al., *Places for People* (Melbourne: Gehl Architects, 2004); J. Gehl et al., *Perth 2009: Public Spaces and Public Life* (Perth: Gehl Architects, 2009).

26. City of Kogarah, "Achieving Sustainability—Kogarah Town Square Development" (report, City of Kogarah [Australia], 2009), www.kogarah.nsw.gov.au/www/html/2075-achieving-sustainability-kogarah-town-square-development.asp).

27. G. P. Metschies, *Prices and Vehicle Taxation* (Germany: Deutsche Geslleschaft fur Technische Zusammenarbeit GmbH, 2001); R. Porter, *Economics at the Wheel: The Costs of Cars and Drivers* (London: Academic Press, 1999).

28. T. Litman, "Changing Vehicle Travel Price Sensitivities: The Rebounding Rebound Effect," *Transportation Policy* (in press, 2012).

29. R. Salzman, "TravelSmart: A Marketing Program Empowers Citizens to Be a Part of the Solution in Improving the Environment," *Mass Transit*, April 2008, 8–11; R. Salzman, "Now That's What I Call Intelligent Transport . . . SmartTravel," *Thinking Highways*, March 2008, 51–58.

30. R. Trubka, P. Newman, and D. Bilsborough, "The Costs of Sprawl," *Environment Design Guidelines* 85 (2010):1–13.

31. C. Ashton-Graham and E. McGregor, in L. Reynolds, ed., *Social Marketing Casebook* (London: Sage, 2011); C. Ashton-Graham, "TravelSmart + TOD = Sustainability and Synergy," in C. Curtis, J. Renne, and L. Bertolini (eds.), *Transit Oriented Development: Making It Happen* (London: Ashgate, 2009); UK Department of Transport, "Smarter Choices—Changing the Way We Travel" (publication of the UK DOT, London, 2004), webarchive.nationalarchives.gov.uk/+ /http://www.dft.gov.uk/pgr/sustainable/smarterchoices/ctwwt/chapter1introduction.pdf; UK Department of Transport, "Making Personalised Travel Planning Work: A Practitioner's Guide" (publication of the UK DOT, London, 2008), www.dft.gov.uk/pgr/sustainable/travelplans/ptp /practitionersguide.pdf.

32. I. Ker, "North Brisbane Household TravelSmart: Peer Review and Evaluation" (publication of the Brisbane City Council, Queensland Transport, and Australian Greenhouse Office, Brisbane, Australia, 2008).

33. Perth Department of Transport, "TravelSmart Household Final Evaluation Report, Murdoch Station Catchment" (report, City of Melville, Socialdata Australia, Department of Transport, Perth 2007).

34. C. Ashton-Graham, "TravelSmart and Living Smart Case Study," *Garnaut Climate Change Review* (publication of the Australian Government, 2008), www.garnautreview.org.au /CA25734E0016A131/WebObj/Casestudy-TravelSmartandLivingSmart-WesternAustralia/%24File /Case%20study%20-%20TravelSmart%20and%20LivingSmart%20-%20Western%20Australia.pdf.

35. D. Wake, "Reduced Car Commuting through Employer-Based Travel Planning in Perth, Australia," *TDM Review* 5, no. 1 (2007): 11–35.

36. R. Garnaut, *The Garnaut Climate Change Review* (Cambridge, UK: Cambridge University Press UK, 2008), 409, www.garnautreview.org.au/pdf/Garnaut_Chapter17.pdf; see also: C. Ashton-Graham, *TravelSmart and Living Smart Case Study* (Cambridge, UK: Cambridge University Press UK, 2008), www.garnautreview.org.au/ca25734e0016a131/WebObj/Casestudy -TravelSmartandLivingSmart-WesternAustralia/$File/Case%20study%20-%20TravelSmart%20 and%20LivingSmart%20-%20Western%20Australia.pdf.

37. B. Davis, T. Dutzik, and P. Baxandall, *Transportation and the New Generation: Why Young*

People Are Driving Less and What It Means for Transportation Policy (publication of the Frontier Group and US Public Interest Research Group, 2012).

38. Telepresence and Videoconferencing Insight Newsletter, September 6, 2010.

39. A. Went, W. James, and P. Newman, "Renewable Transport" (CUSP Discussion Paper, Curtin University Sustainable Policy Institute, Perth, Western Australia, January 2008), www .sustainability.curtin.edu.au/renewabletransport.

40. A. Simpson, "The Electric Revolution Is on Track," *Business Spectator*, September 1, 2009; A. Simpson, "Environmental Attributes of Electric Vehicles in Australia" (CUSP Discussion Paper, Curtin University Sustainable Policy Institute, Perth, Western Australia, 2009), www.sustainability.com.au/renewabletransport.

41. M. Kintner-Meyer, K. Schneider, and R. Pratt, *Impacts Assessment of Plug-in Hybrid Vehicles on Electric Utilities and Regional US Power Grids, Pt. 1: Technical Analysis* (report, Pacific Northwest National Laboratory, US Department of Energy, DE-AC05–76RL01830, 2007).

42. Al Gore, *A Generational Challenge to Repower America*, Speech to Daughters of the American Revolution, Constitution Hall, Washington, DC, July 17, 2008, www.ens-newswire.com.

43. L. Mastny, ed., *Biofuels for Transportation* (publication of the Worldwatch Institute, GTZ, German Ministry of Agriculture, Washington, DC, 2006).

44. D. Demerjian, "As Fuel Costs Rise Airlines Can't Make the Math Work," *Autopia*, June 10, 2008, blog.wired.com/cars/2008/06/prices-for-jet.html, accessed June 25, 2009; R. Seaney, "High Oil Prices Spell Disaster for Airlines," ABC News Internet Ventures, 2009, a.abcnews .com/m/screen?id=4847008&pid=74, accessed August 25, 2009.

45. Gilbert and Perl, *Transport Revolutions*.

46. International Energy Agency, "Technology Roadmap—Biofuels for Transport" (IEA report, Paris, 2011).

47. D. Bradbury, "Airships Float Back to the Future," BusinessGreen.com, September 2, 2008.

48. C. Creutz and E. Fujita, "Carbon Dioxide as a Feedstock," *Carbon Management: Implications for R&D in Chemical Sciences* (report of the Commission on Physical Sciences, Mathematics, and Applications [CPSMA], National Research Council, Washington, DC, 2001).

49. Jared Diamond, *Collapse: How Societies Choose to Fail or Succeed* (New York: Viking Books, 2009).

50. D. Lankshear and N. Cameron, "Peak Oil: A Christian Response," *Zadok Perspectives* 88 (2005): 9–11.

The Pent-Up Demand for Transit-Oriented Development and Its Role in Reducing Oil Dependence

13

JOHN L. RENNE

Contributors to this book have discussed America's oil dependence and have offered some creative solutions that focus on making transportation systems more sustainable and oil-free. However, as some authors in the book have noted, transportation-only solutions are not enough; we also need to create communities that facilitate transit use, walking, and bicycling through the design of the built environment. This chapter focuses on the role of transit-oriented development (TOD) in reducing oil dependence.

The chapter summarizes the massive pent-up demand for TOD across the United States and the significant opportunity for infill development in existing rail station precincts. This chapter shows that regions have a disproportionately higher share of jobs in rail station precincts in comparison to a dearth of housing in these same areas. This stems from rail systems that serve large concentrations of jobs in downtown areas with inefficient, low-density, auto-dominated land uses around suburban stations. The dominant model of development around rail stations in America has been park-and-ride; however, most TODs have been successful in blending commuter parking with residential and commercial uses, thus creating a more efficient pattern of land use around rail stations. The analysis presented below examines the average number of people per passenger rail station precinct, by region. It presents the percentage of stations, by region, that meet minimum-density thresholds of 8, 15, and 25 gross units per acre, and also calculates the number of people per region who could be accommodated in rail station precincts through infill development to meet these minimum thresholds. This infill development would result from the growth of the next 100 million Americans, who are forecasted to arrive by 2050.

A national TOD strategy could encourage new housing in station precincts, which would seek to better utilize existing infrastructure to accommodate future population growth. In calculating how to absorb the next 100 million Americans, a model is shown that first builds out all existing rail stations (at 8 and 15 units per acre) and then shows how many new stations would be needed in America to accommodate this future growth, based on varying percentages of the market that would want to live in TODs. It also calculates a rough estimate of how much the national yearly investment would be from 2015 to 2050 to accomplish this growth potential. Finally, the chapter provides rough forecasts of the oil-saving benefits of a national TOD strategy based on recent studies that have found that TOD residents yield significantly fewer vehicle-miles traveled (VMTs) as compared with their suburban counterparts. In short, America could save as much as 942 million barrels of oil per year by 2050 if an ambitious target of 90 percent of future population growth occurred in TODs. If only 30 percent of future growth is targeted to TODs, a national savings of 485 million barrels of oil per year could be achieved. As a reference point, the US Strategic Petroleum Reserve includes a one-time supply of 726 million barrels; thus such savings due to TOD, occurring on a yearly basis, could be very significant to the United States for a number of reasons.[1]

While some may argue that directing future populations to live in TODs might be the vision of environmentalists and/or heavy-handed planners, market evidence is proving the contrary. Recent trends point to increased demand for a TOD product. For example, a recent study found that 81 percent of Generation Y (those Americans born between 1983 and 2001), also referred to as Echo Boomers, indicated that it was very or somewhat important to live near transit, and 67 percent would pay a premium for this amenity.[2] When older Americans seeking smaller housing units are included, the potential market for TOD is even more significant. In *The Option of Urbanism*, Christopher Leinberger discusses the need for more walkable communities in America in order to satisfy the unmet demand from both the Baby Boomer and Echo Boomer generations.[3] He finds that 40 percent or more of the entire nation would want to live in a walkable neighborhood with transit access if such a product existed in the marketplace—today.

Richard Florida's *The Great Reset* indicates that America's Great Recession of 2008 was due in large part to a crisis of confidence in our current system—based largely on access to cheap oil that fuels sprawl and automobile dependence.[4] America's recovery from the Great Recession lies in connecting people, jobs, and goods from the global economy to the local sidewalk. Arthur Nelson notes that more than 50 percent of the buildings that will exist in America in 2030 have not yet been built.[5] Just as economic investment flourished in the 1950s by building suburbia around highway interchanges, billions of investment dollars could be directed to TODs over the next half century. This chapter illustrates that the next real estate engine could be TOD—driven by market preferences for pent-up demand and population growth, not by planners trying to control the market.

Welcome to the New 1950s

America in the 1950s was a nation on the brink of significant economic prosperity but it was not until President Eisenhower signed the 1956 Highway Bill, creating new transportation infrastructure, that a tidal wave of investment was unleashed, building suburban America. The suburban-focused paradigm for real estate investment ended in 2008 as the market crashed. Millions of homeowners and investors lost significant wealth that is unlikely to return to previous levels anytime soon, but this was not the case in most walkable TOD communities, where land retained more value during the downturn.[6]

Today, major investors and companies are shifting their focus to TOD. Our nation has reached the end of the Suburban Era, and the Era of TOD and Walkable Communities is quickly becoming the next dominant paradigm for real estate in America. Moreover, our infrastructure base of 4,416 existing train stations can be used to facilitate this new TOD era, since the vast majority of them are underutilized and are surrounded by low-density land uses. These stations are ripe for redevelopment and have caught the attention of developers and investors. For example, Forest City, a $10.5 billion New York Stock Exchange company that made fortunes building suburbia over the past 50 years, is now shifting a significant share of their future portfolio of new construction to TOD.[7] *Emerging Trends in Real Estate*, an annual publication by the Urban Land Institute and PricewaterhouseCoopers, has rated TOD as the best risk-adjusted investment in real estate every year since 2005, as it appreciates faster in up-markets and holds value better in down-markets.[8]

Each year's data increasingly supports this hypothesis, and savvy investors are repositioning their portfolios to take advantage of TOD. Soon, as TOD becomes even more mainstream, a large pool of capital will be shifted off the sidelines into this emerging class of real estate. These investments will support the redevelopment of land around a national network of underutilized rail stations while simultaneously winning political champions to expand infrastructure that will create new development opportunities. The expansion of freeways and interchanges, which fueled suburban development into the twenty-first century, provided 50 years of stability for real estate investors. That era is over and will be replaced by the Era of TOD and Walkable Communities. Based on this new paradigm, what does it mean for future scenarios of oil consumption? Before we can forecast the answer to that question, we must examine the existing supply of infrastructure to serve this emerging market.

A Nation of Nearly-Empty Rail Station Precincts

Over the past two centuries, our nation has assembled a tremendous asset that was overlooked during the Era of Suburbia—rail stations. In 2000, 6.18 percent of Americans lived within a half mile of one of these stations; however, by 2010, the end of the decade in which the Era of Suburbia gave its last gasp before crashing, only 5.78 percent of Americans lived near a station.

While there is some debate about minimum densities to support transit ridership, most planners agree that higher densities result in larger shares of ridership. Best practices indicate that minimum gross densities for successful rail-based TODs should be at least 15 units per acre, but some even advocate minimum densities of as much as 30 units per acre.[9] The reality is that most rail precincts in America fail to achieve 8 units per acre—just half the density of the minimum recommended density. A gross density of 8 units per acre within a half-mile precinct of a rail station results in 4,000 housing units, while a minimum density of 15 units per acre includes 7,500 units. At 25 units per acre, a rail precinct would contain 12,500 units. Table 13.1 shows the percentage of all rail station precincts in the United States, based on minimum housing density thresholds. Only 38 percent of stations have a minimum gross density of 8 units per acre while 20 percent of stations in America meet the 15 units per acre minimum threshold and 11 percent meet the 25 units per acre target.

A Regional Snapshot of People and Jobs in Rail Precincts

Of all regions across the country, transit usage is the highest in the New York region, which accounts for approximately 35 percent of all transit ridership in the United States.[10] The New York region also contains 22 percent of all rail stations in the United States and the region therefore skews national averages, since densities around rail stations in this region are much higher than the rest of the country (58 percent, 49 percent, and 38 percent of stations with minimum densities above 8 units, 15 units, and 25 units per acre, respectively). Forty-two percent (16 of 38) of all regions with rail stations have 0 percent of stations that meet the minimum density threshold of 8 units per acre. Seventy-one percent (27 of 38) of all regions have 0 percent of stations that

Table 13.1 Population density and availability of rail stations in 39 US cities with passenger rail

Region	Percent of stations with minimum household density across half-mile-radius station precinct		
	8 units per acre (4,000 units within precinct)	15 units per acre (7,500 units within precinct)	25 units per acre (12,500 units within precinct)
Albuquerque (13 stations)	0%	0%	0%
Atlanta (41 stations)	5%	0%	0%
Austin (9 stations)	0%	0%	0%
Baltimore (67 stations)	28%	4%	0%
Boston (325 stations)	39%	20%	4%
Buffalo (16 stations)	0%	0%	0%
Charlotte (15 stations)	0%	0%	0%
Chicago (417 stations)	30%	13%	4%
Cleveland (90 stations)	2%	0%	0%
Dallas (94 stations)	21%	0%	0%
Denver (54 stations)	38%	0%	0%
Detroit (12 stations)	0%	0%	0%
Eugene (28 stations)	0%	0%	0%
Harrisburg (5 stations)	0%	0%	0%
Houston (16 stations)	0%	0%	0%
Jacksonville (8 stations)	0%	0%	0%
Kansas City (55 stations)	0%	0%	0%
Las Vegas (54 stations)	4%	0%	0%
Little Rock (13 stations)	0%	0%	0%
Los Angeles (151 stations)	19%	9%	3%
Memphis (23 stations)	0%	0%	0%
Miami (67 stations)	37%	15%	4%
Minneapolis & St. Paul (25 stations)	12%	0%	0%
Nashville (6 stations)	0%	0%	0%
New Orleans (97 stations)	15%	0%	0%
New York (951 stations)	58%	49%	38%
Norfolk (14 stations)	0%	0%	0%
Philadelphia (610 stations)	45%	10%	2%
Phoenix (32 stations)	3%	0%	0%
Pittsburgh (86 stations)	1%	0%	0%
Portland (141 stations)	42%	17%	0%
Sacramento (61 stations)	6%	0%	0%
Salt Lake City (48 stations)	2%	0%	0%
San Diego (81 stations)	17%	7%	0%
San Francisco (424 stations)	57%	34%	20%
Seattle (72 stations)	19%	8%	0%
St. Louis (37 stations)	0%	0%	0%
Tampa (11 stations)	0%	0%	0%
Washington, DC (131 stations)	24%	9%	2%
All Stations (4,416 Stations)	36%	20%	11%

meet the minimum density threshold of 15 units per acre, and 82 percent (31 of 38) of all regions have 0 percent of stations that meet the minimum density threshold of 25 units per acre. America needs to increase housing densities significantly around 80–90 percent of its stations just to meet the minimum density that is necessary to support transit ridership. Otherwise, we are not utilizing the stations for their intended purpose and we're allowing station precincts to be low-density and auto-centric—a waste of rail infrastructure, which only leads to lower ridership and higher subsidies to support transit systems. This topic will be addressed later in the chapter.

In New York, an average of 8,313 people live within the half-mile-radius station precinct of the region's 951 rail stations, which accounts for 36 percent of the entire metropolitan population (see figs. 13.1 and 13.2). Four regions contain 15–25 percent of the regional population within transit precincts. These include Philadelphia, Chicago, Boston, and San Francisco. However, in looking at the average number of people per station, Los Angeles out-performs all others except New York, with an average of 6,639 people per rail precinct. However, the data also shows that only 9 percent of stations in Los Angeles exceed the 15 units per acre minimum threshold and only 2 percent of stations exceed the 25 units per acre minimum threshold. In comparison, the San Francisco region has an average of only 2,818 people per station, but 34 percent of stations exceed the 15 units per acre minimum threshold and 20 percent of stations exceed the 25 units per acre minimum threshold. Los Angeles can be characterized as "crowded but not dense enough" for most of its rail precincts. The issue for San Francisco is different, as it has a considerable percentage of stations that meet minimum density targets, but it has a long way to go for the remaining stations and is more similar to New York in this regard.

Other cities, such as Portland, Washington, DC, Eugene, San Diego, New Orleans, Pittsburgh, Baltimore, Las Vegas, Salt Lake City, Cleveland, Los Angeles, and Sacramento all have between 5 and 10 percent of the metropolitan population living within station precincts, whereas less than 5 percent of the population in the remaining 22 regions live within walking distance of a rail station.

In looking at regional employment, New York again tops the list, with 46 percent of jobs across the region accessible by rail. San Francisco and Chicago are at about 35 percent, and Philadelphia, Portland, and Eugene each have about 30 percent of all jobs accessible by rail. Two-thirds of all regions have between 10 and 20 percent of all regional jobs accessible by rail (see fig. 13.3).

In light of the lack of housing density at most rail stations and the minimal share of regional population within rail precincts, the market for new housing in transit precincts appears to be ripe across the United States because there is a disproportionate

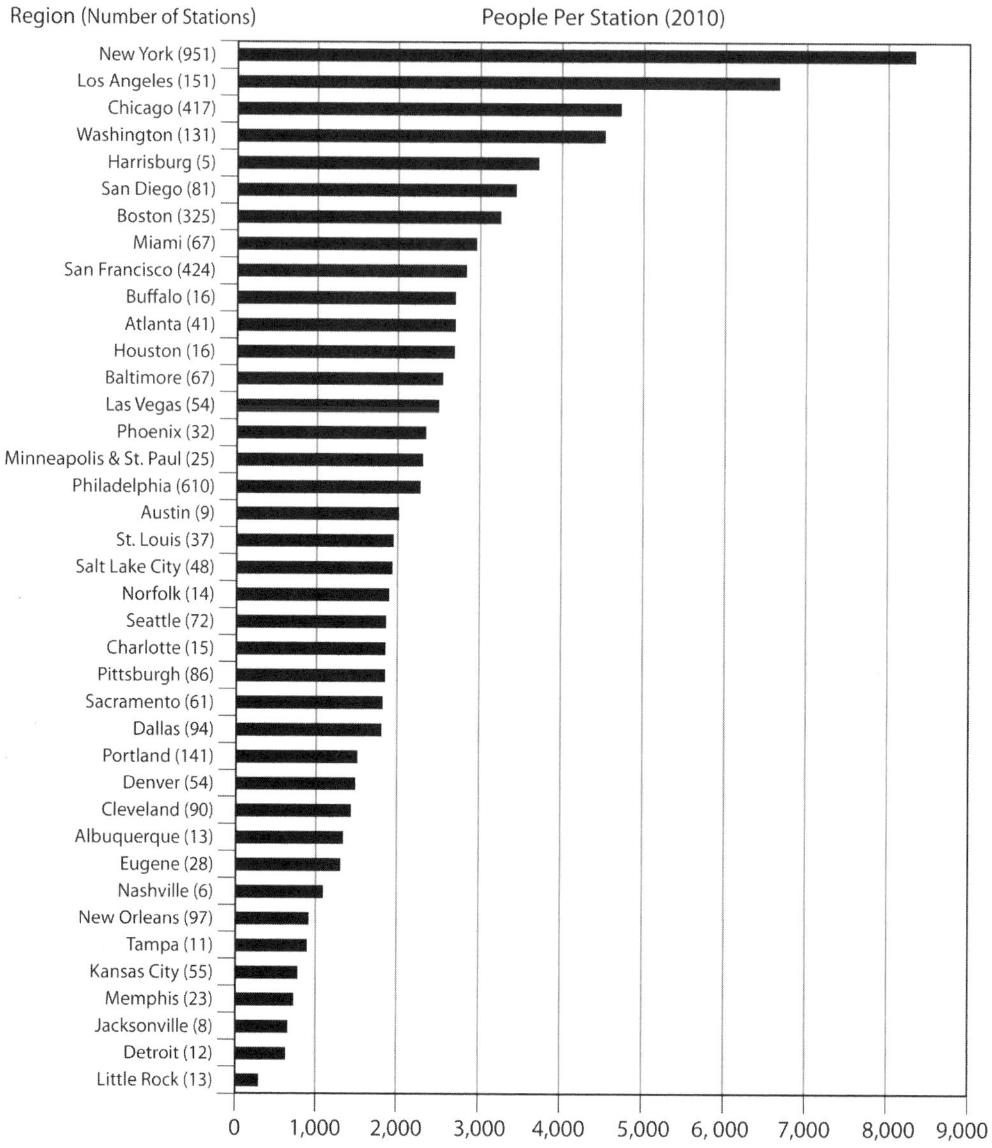

Region (Number of Stations)   People Per Station (2010)

Figure 13.1   Average number of people per station precinct by region in 2010. (Source: US Census and the National TOD Database.)

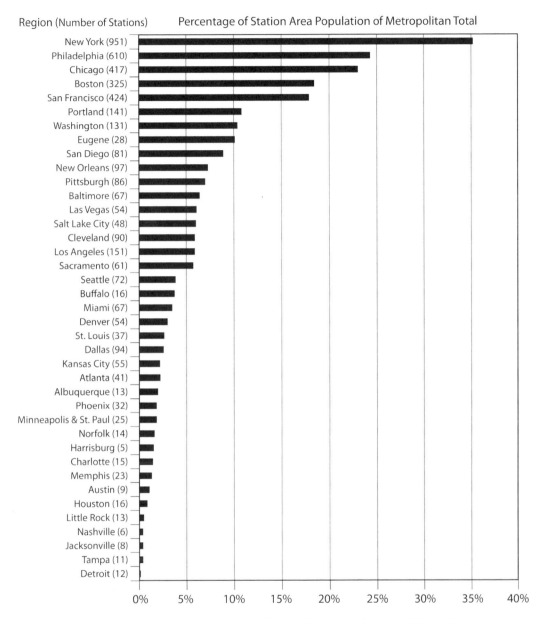

Figure 13.2   Percentage of station-area population of metropolitan total by region in 2010. (Source: US Census and the National TOD Database.)

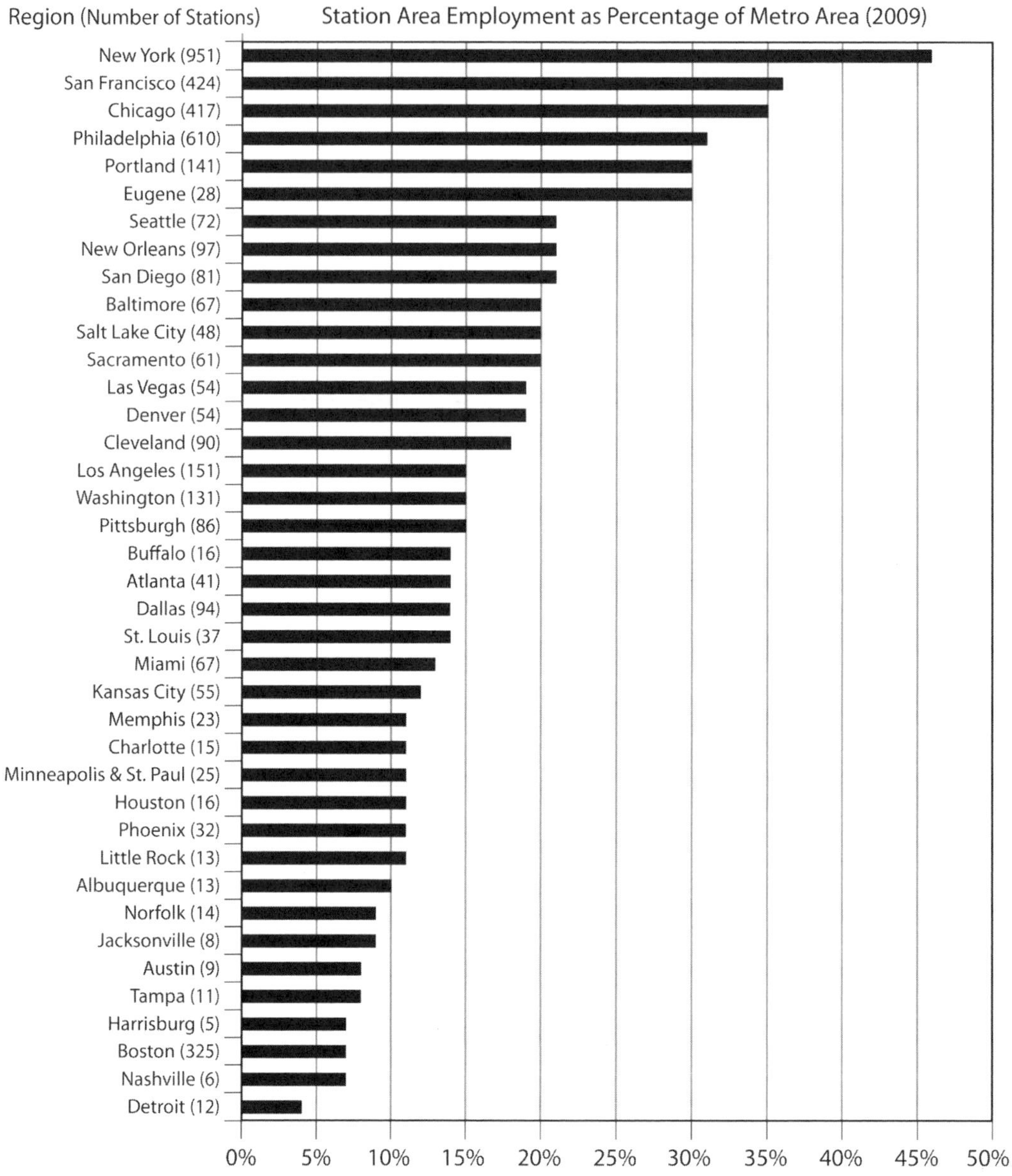

Region (Number of Stations)   Station Area Employment as Percentage of Metro Area (2009)

Figure 13.3   Station-area employment as percentage of metro area in 2009. (Source: Local Employment Dynamics and the National TOD Database.)

number of jobs in rail precincts as opposed to people actually living there. This is likely due to most transit agencies' focus over the past few decades on park-and-ride. However, in thinking about accommodating the next 100 million Americans by 2050, a national TOD strategy that encourages structured parking for auto-commuters to be supplanted by new housing in rail precincts seems not just logical but necessary, due to the strong demand for TOD housing, as noted above, and the lack of housing in comparison with jobs.

The remainder of this chapter will focus on establishing national targets for housing in rail precincts. This does not mean that creating new workplaces, services, retail outlets, and other land uses is not important for the success of TOD in America—they are all very important. In fact, a recent meta-analysis of travel and the built environment found that seven "Ds" are important in creating places that facilitate walking and transit riding. These include (1) Density, (2) Diversity (land-use mix), (3) Design, (4) Destination accessibility, (5) Distance to transit, (6) Demand management, and (7) Demographics.[11] However, our national population is projected to grow significantly and our roadway system has reached capacity in many regions. Infill development of housing near rail stations, and the construction of new rail infrastructure to meet growing demand for this product-type, could not only connect people to jobs, but do so in a manner that minimizes oil use.

How Much New Housing Can Existing Station Precincts Accommodate?

With demand for TOD nearly 14 times greater than supply across the nation, it quickly becomes apparent that there is an immediate market opportunity for infill housing near existing stations. A key question is how much infill housing can be accommodated without having to build new infrastructure? The answer to this lies in how much density is ultimately legally allowable (i.e., potential supply for new TOD housing) in rail precincts and how this relates to market demand. Calculating legally allowable density is nearly impossible, given the disparate nature of rail stations across thousands of local jurisdictions, and few studies have previously examined the total population living near rail stations. Table 13.2 presents the total population living in all rail precincts, by region, and projects the additional number of people that could be accommodated in each region at existing stations, based on minimum gross densities of 8 units, 15 units, and 25 units per acre across all stations.

Based on a projected 2050 population of 409 million Americans (100 million new residents more than in 2010), existing stations could accommodate 10.8 percent, 20.3 percent, or 32.4 percent of the entire population based on the level of minimum density achieved. By 2050 Generation Y, currently with strong desire for TOD, will be

Table 13.2 Potential for rail-accessible infill development in 39 US cities with passenger rail

Region	2010 Population	Number of additional people who could be accommodated in rail-station precincts through infill development		
		10,000 people per station target (8 units per acre)	18,750 people per station target (15 units per acre)	30,000 people per station target (25 units per acre)
Albuquerque (13 stations)	17,716	112,284	226,034	372,284
Atlanta (41 stations)	110,568	299,432	658,182	1,119,432
Austin (9 stations)	18,023	71,977	150,727	251,977
Baltimore (67 stations)	172,643	497,357	1,083,607	1,837,357
Boston (325 stations)	1,055,742	2,194,258	5,038,008	8,694,258
Buffalo (16 stations)	43,493	116,507	256,507	436,507
Charlotte (15 stations)	28,109	121,891	253,141	421,891
Chicago (417 stations)	1,972,937	2,197,063	5,845,813	10,537,063
Cleveland (90 stations)	128,486	771,514	1,559,014	2,571,514
Dallas (94 stations)	169,901	770,099	1,592,599	2,650,099
Denver (54 stations)	82,879	457,121	929,621	1,537,121
Detroit (12 stations)	7,363	112,637	217,637	352,637
Eugene (28 stations)	35,646	244,354	489,354	804,354
Harrisburg (5 stations)	18,546	31,454	75,204	131,454
Houston (16 stations)	43,100	116,900	256,900	436,900
Jacksonville (8 stations)	4,952	75,048	145,048	235,048
Kansas City (55 stations)	43,616	506,384	987,634	1,606,384
Las Vegas (54 stations)	134,822	405,178	877,678	1,485,178
Little Rock (13 stations)	3,422	126,578	240,328	386,578
Los Angeles (151 stations)	1,002,530	507,470	1,828,720	3,527,470
Memphis (23 stations)	16,657	213,343	414,593	673,343
Miami (67 stations)	195,740	474,260	1,060,510	1,814,260
Minneapolis & St. Paul (25 stations)	57,616	192,384	411,134	692,384
Nashville (6 stations)	6,520	53,480	105,980	173,480
New Orleans (97 stations)	87,244	882,756	1,731,506	2,822,756
New York (951 stations)	7,905,769	1,604,231	9,925,481	20,624,231
Norfolk (14 stations)	26,550	113,450	235,950	393,450
Philadelphia (610 stations)	1,394,635	4,705,365	10,042,865	16,905,365
Phoenix (32 stations)	76,012	243,988	523,988	883,988
Pittsburgh (86 stations)	159,051	700,949	1,453,449	2,420,949
Portland (141 stations)	239,249	1,170,751	2,404,501	3,990,751
Sacramento (61 stations)	110,887	499,113	1,032,863	1,719,113
Salt Lake City (48 stations)	93,377	386,623	806,623	1,346,623
San Diego (81 stations)	276,102	533,898	1,242,648	2,153,898
San Francisco (424 stations)	1,194,652	3,045,348	6,755,348	11,525,348
Seattle (72 stations)	135,668	584,332	1,214,332	2,024,332
St. Louis (37 stations)	73,427	296,573	620,323	1,036,573
Tampa (11 stations)	9,813	100,187	196,437	320,187
Washington, D.C. (131 stations)	601,102	708,898	1,855,148	3,328,898
All stations (4,416 stations)	17,754,565	26,405,435	65,045,435	114,725,435
Total U.S. population in 2050	408,745,538	10.8%	20.3%	32.4%

well into their retirement years. Unfortunately, if current demand levels persist at 80 percent of all Americans, our best hope for TOD infill development, at 25 units per acre, is for new housing supply to accommodate just 114.7 million Americans. To meet the full market demand, as discussed below, many more stations will have to be built.

Building out all rail stations in America at a minimum density of 25 units per acre is likely to produce significant local opposition among NIMBY communities.[12] In *Zoned Out*, Jonathan Levine discusses NIMBYism and how the demand for mixed-use and dense communities is significantly constrained across America due to restrictive local zoning policies that maintain low-density, auto-oriented land uses.[13] In most regions, single-family neighbors feel that 8 units per acre is well beyond acceptable density levels, as most housing developments measure only 1–4 units per acre, maximum. When people romanticize the American Dream, a quarter-acre lot is often a reference point. Gross densities are significantly lower when accounting for public space.

Even if America could manage to build out around stations to achieve a gross density of 8 units per acre by 2050, accommodating 26.4 million of the 409 million forecasted population, demand for TOD housing would remain over 12 times greater than the supply, assuming demand remains at 80 percent. If acceptance for higher-density living in TODs is truly as high as reported for Generation Y by the National Association of Realtors, then NIMBY resistance is likely to fade over the next 10–20 years, resulting in stronger political support for TOD as Generation Y becomes the emerging decision-making generation in America.

How Many New Stations Are Needed to Accommodate the Next 100 Million Americans?

As calculated in the previous section, existing rail stations cannot accommodate the housing demand for the next 100 million Americans in TODs, except for the scenario where minimum gross densities are 25 units per acre. As shown in table 13.2, this level of density could accommodate nearly 115 million Americans; however, housing an average of 30,000 people per precinct at every station in America seems very unrealistic, given that such densities are so much disproportionally greater than existing suburban densities. Therefore, a key follow-up question is: How many new stations are needed to accommodate the next 100 million Americans, assuming that all existing stations are built out to 8 units or 15 units per acre?

Table 13.3 shows that at 8 units per acre, 359 new stations would be needed to accommodate 30 percent of the next 100 million Americans, while 7,359 new stations would be needed to accommodate the entire growth of America's expected 2050 population in TODs.[14] Having 100 percent of future growth locate in TODs is unrealistic,

Table 13.3 Projected need for rail station investment in US cities, 2015–2050

Percentage of the next 100 million Americans to live in station precincts	Number of new stations needed for an average of 10,000 people per station/ 8 units per acre[a]	Total investment, 2015–2050 (in billions)[b]	Number of new stations needed for an average of 18,750 people per station/ 15 units per acre[a]	Total investment, 2015–2050 (in billions)[b]
30%	359	$18.0	—	$0.0
40%	1,359	$68.0	—	$0.0
50%	2,359	$118.0	—	$0.0
60%	3,359	$168.0	—	$0.0
70%	4,359	$218.0	264	$13.2
80%	5,359	$268.0	798	$39.9
90%	6,359	$318.0	1,331	$66.5
100%	7,359	$368.0	1,864	$93.2

[a] Assumes all existing stations are also built out to the target densities.

[b] Assumes the total cost for entire infrastructure, on average, is $50 million per station, in 2012 dollars.

but table 13.3 provides the data at various levels for illustrative purposes. By increasing average gross densities to 15 units per acre per station, only 1,864 new stations would need to be constructed to accommodate the entire population growth by 2050. The table also projects the cost of constructing the new rail infrastructure. If we increase densities to 15 units per acre per station, the nation could accommodate over 60 percent of the projected population growth in TODs without having to build a single new station in the country. However, to accommodate 60 percent of the population at 8 units per acre, the nation would need to construct 3,359 new stations at an estimated cost of $4.8 billion per year from 2015 to 2050.[15] In revisiting the demand estimate of 80 percent of Generation Y desiring to live in a TOD setting, such growth could be accommodated at a cost of $7.7 billion per year to construct 5,359 new stations at 8 units per acre per station, or $1.1 billion per year to build 798 new stations at 15 units per acre per station.

Oil-Saving Benefits of a National TOD Strategy

The final goal of this chapter is to broadly estimate the oil-saving benefits of a national TOD strategy. A recent national study found, on average, that TOD residents traveled 50 percent fewer vehicle miles (VMT) than typical suburbanites in conventional auto-centric communities.[16] Research from several studies finds that VMT decreases by approximately 20–30 percent every time density is doubled. Some studies find that density is a proxy variable for some of the other seven Ds noted above.[17] Nevertheless, in order to estimate the oil-saving benefits of a national TOD strategy, developing such assumptions is necessary.

In 2010, 308.7 million Americans consumed 4.385 billion barrels of petroleum within the transportation sector.[18] As shown in the introductory chapter, petroleum consumption in the United States increased each year since 1949 until it fell in 2008 and 2009, followed by an uptick in 2010. To develop a per-capita rate of petroleum consumption, data were averaged from 1990 to 2010, resulting in an average per-capita petroleum consumption of 15 barrels per person per year for transportation purposes (note that this estimate includes both passenger and goods movements). Applying this consumption rate forward to the next 100 million Americans results in a baseline annual consumption of 6.151 billion barrels by 2050, which represents a rate of 40 percent growth from 2010.

In projecting future TOD oil-consumption savings, two scenarios are presented; each forecast is rudimentary and is based on TOD residents driving half as many VMTs, and therefore consuming half as much petroleum as compared with conventional suburban residents. If 30 percent of the next 100 million Americans become TOD residents, total annual petroleum consumption in America would fall to 5.67 billion barrels (from projected 6.151 billion barrels); whereas if 90 percent of the next 100 million Americans move into TODs, consumption would be 5.217 billion barrels per year. Therefore, if we hold everything constant from today and forecast the impact of more people living in TODs, which is characterized by half the driving of conventional neighborhoods, the nation could reduce the growth in oil consumption from 40 percent to 29 percent, if just 30 percent of the future growth occurs in TODs. This represents a savings of 485 million barrels of oil per year as compared with the baseline. If we set an ambitious target of 90 percent of future population growth in TODs, our nation would be able to reduce growth in petroleum consumption down to 19 percent, resenting a savings of 942 million barrels of oil per year by 2050 as compared with the baseline forecasts. Again, this calculation is purely for illustrative purposes—reality will vary depending on a myriad of variables. However, the key point is that if we encourage people to live in TODs, substantial reductions in the growth of oil consumption are attainable, due entirely to fewer VMTs.

A Market-Driven Policy Path Forward for the United States

This chapter has attempted to forecast the impact of TODs on reducing oil dependence among the next 100 million Americans by 2050. Any method of forecasting future scenarios is subject to criticism, but the intention of this exercise is to establish a basis for creating policy today that can result in a more sustainable tomorrow. A market-driven policy path forward is needed. As discussed above, demand for living in TODs across America by Generation Y is approximately 14 times greater than existing supply. Even if demand were only twice as much as supply, a significant market opportunity would exist to build new housing in TODs. The sort of pent-up demand due to limited supplies of TOD housing could create a new paradigm in the real estate sector that could fuel trillions of dollars in private investment in TODs over the next 40–50 years across the United States. This titanic shift in private investment has already begun as properties near rail stations have held their values, if not increased in value, during the worst economic downturn since the Great Depression. Such a shift in real estate investment is similar to the suburban boom, which began in the 1950s and continued strong until the mid–2000s, a period of approximately the same length.

Pent-up demand is a good thing for an emerging product, such as TOD; however, current policies in America subsidize low-density residential development over TOD, thus creating a non-level playing field. For example, in a sample of over 100 TODs across the United States, 63 percent of the residents were renters as compared with a rate of 38 percent of renters across the same metro regions.[19] As people seek the tax benefits of mortgage interest deduction, they often do not have the option to afford to buy in TODs and are literally driven to buy in the suburbs over renting in TODs.[20] Market-driven policies would reduce subsidies to the suburbs. Higher per-unit costs of infrastructure to taxpayers in low-density neighborhoods as compared with denser TODs is further evidence that our current polices are *not* market-oriented. Market-oriented policies would not result in massive pent-up demand for TOD while so many Americas are forced to choose a neighborhood type that they do not consider ideal.

Critics often point to large subsidies for infrastructure in TODs; however, they often ignore that these same subsidies are present in most non-TOD developments. For example, critics will include the cost of rail construction, parks, sidewalks, and utilities within the cost estimates for TOD but fail to remember that such costs are present in non-TOD developments as well. Highway interchanges and roads leading to conventional subdivisions, which enabled these communities to be constructed in the first place, are not insignificant costs. Moreover, tax increment financing (TIF) and tax credits, such as the New Market Tax Credit and the Low Income Housing Tax Credit sometimes used in TODs, are not exclusive to TOD. Many, many projects that are not located in TODs have been financed with TIF and tax credits.

There is a large and growing body of literature about the economic, environmental, social, and public-health benefits of development that is walkable, bikeable, and transit accessible, such as TOD.[21] Building upon that literature, this chapter provides a sobering reality. Unless we are willing as a nation to significantly increase densities around existing and future rail stations, TOD will remain a niche product and pent-up demand will far outpace supply, benefiting just a small segment of the population that can afford to live in TODs.

TOD represents the perfect opportunity for bipartisan support because it creates new jobs, both construction and permanent, due to its mixed-use nature, which can help to breathe new life into the moribund national market. The roles for local, state, and federal governments are relatively simple. First, local government needs to re-move barriers that force rail precincts to be low density and auto dominated. Second, state and federal governments should create a level playing field by stopping subsidies to sprawl. This could entail cutting budgets for massive highway projects, the mort-gage interest deduction, and other programs that cost taxpayers trillions. Finally, for every tax dollar saved, some of this savings should be reinvested into building new infrastructure such as new rail lines, local streets and sidewalks, bicycle lanes, and parking structures to blend commuter parking in TODs. Not only will the savings of ending subsidies to sprawl help to balance our national budget, but the reinvestment of a portion of this into rail and TOD will help to leverage new investment, create new jobs and housing to meet pent-up demand, and help the United States to create a prosperous twenty-first century.[22]

Notes

1. US Department of Energy, "Strategic Petroleum Reserve" (US DOE report, 2012), www.fossil.energy.gov/programs/reserves/, accessed July 23, 2012.

2. Brad Broberg, "Generation Y: The Future Generation of Home Buyers," *On Common Ground* (publication of the National Association of Realtors, 2010), 4–9, www.realtor.org/publications/on-common-ground/summer–2010-megatrends-for-the-decade.

3. Christopher Leinberger, *The Option of Urbanism: Investing in a New American Dream* (Washington, DC: Island Press, 2009).

4. Richard Florida, *The Great Reset: How the Post-Crash Economy Will Change the Way We Live and Work* (New York: Harper, 2011); Peter Newman, Timothy Beatley, and Heather Boyer, *Resilient Cities: Responding to Peak Oil and Climate Change* (Washington, DC: Island Press, 2009).

5. Arthur Nelson, *Reshaping Metropolitan America: Development Trends and Opportunities to 2030* (Washington, DC: Island Press, 2012).

6. Christopher Leinberger, "Now Coveted: A Walkable Convenient Place," *New York Times*, May 25, 2012, www.nytimes.com/2012/05/27/opinion/sunday/now-coveted-a-walkable-con venient-place.html?_r=1, accessed July 20, 2012.

7. Personal communication.

8. Jonathan Miller, *Emerging Trends in Real Estate* (annual report of the Urban Land Institute and PricewaterhouseCoopers, Washington, DC, 2005, 2006, 2007, 2008, 2009, 2010, 2011, and 2012).

9. Greater Cleveland Regional Transit Authority, *Transit Oriented Development Guidelines* (publication prepared by TRA, Nelson Nygaard, and Van Auken Atkins, 2007), www.riderta .com/tod/guidelines/#density, accessed July 20, 2012.

10. John Pucher, "Renaissance of Public Transport in the United States?" *Transportation Quarterly* 56, no. 1 (2002): 33–49, ejb.rutgers.edu/faculty/pucher/TQPDF.pdf, accessed July 20, 2012.

11. Reid Ewing and Robert Cervero, "Travel and the Built Environment," *Journal of the American Planning Association* 76, no. 3 (2010): 1–30, reconnectingamerica.org/assets/Uploads /travelbuiltenvironment20100511.pdf, accessed July 20, 2012.

12. NIMBY stands for Not In My BackYard.

13. Jonathan Levine, *Zoned Out: Regulation, Markets, and Choices in Transportation and Metropolitan Land Use* (Washington, DC: Resources for the Future, 2006).

14. Stations that are planned or under construction, but not operational according to the 2010 National TOD Database, would be included among these new stations.

15. These cost estimates only include rail infrastructure. Certainly, the higher the density of the TOD, the greater the required increase in local and private investment; however, such costs are not typically covered by the federal government. Still, greater densities reduce the federal burden to fund the construction of many stations serving low-density communities, so cost savings to the federal government in not having to build additional stations could be reallocated in order to provide funds for local infrastructure needs associated with higher-density settings.

16. G. B. Arrington and Robert Cervero, "Effects of TOD on Housing, Parking, and Travel" (Transit Cooperative Research Program Report 128, Transportation Research Board of the National Academies, Washington, DC, 2008), onlinepubs.trb.org/onlinepubs/tcrp/tcrp_rpt_128 .pdf, accessed July 20, 2012.

17. Ewing and Cervero, "Travel and the Built Environment"; see also: John Holtzclaw, Robert Clear, Hank Dittmar, Hank Goldstein, and Peter Haas, "Location Efficiency: Neighborhood and Socio-Economic Characteristics Determine Auto Ownership and Use—Studies in Chicago, Los Angeles and San Francisco," *Transportation Planning and Technology* 25, no. 1 (2002): 1–25, www.tandfonline.com/doi/abs/10.1080/03081060290032033, accessed July 20, 2012.

18. US Energy Information Administration, *Annual Energy Review* (EIA publication, US DOE, 2011), www.eia.gov/totalenergy/data/annual/index.cfm, accessed July 20, 2012.

19. John Renne, "Transit-Oriented Development: Measuring Benefits, Analyzing Trends and Evaluating Policy" (doctoral dissertation, Edward J. Bloustein School of Planning and Public Policy, Graduate School of New Brunswick, Rutgers University, New Brunswick, NJ, 2005).

20. A growing body of literature exists on TODs and affordability. See, for example: "Maintaining Diversity in America's Transit-Rich Neighborhoods: Tools for Equitable Neighborhood

Change" (report, Dukakis Center for Urban and Regional Policy, Northeastern University, Boston, MA, 2010), www.dukakiscenter.org/tod/, accessed July 20, 2012.

21. See: Peter Newman and Jeffrey Kenworthy, *Sustainability and Cities: Overcoming Automobile Dependence* (Washington, DC: Island Press, 1999); Howard Frumkin, Lawrence Frank, and Richard Jackson, *Urban Sprawl and Public Health: Designing, Planning and Building for Healthy Communities* (Washington, DC: Island Press, 2004).

22. *Information Resources*:

Data in this chapter were provided by the National TOD Database, a project of the Center for Transit-Oriented Development. The database and following information can be accessed at: toddata.cnt.org/. Intended as a tool for planners, developers, government officials, and academics, the Database provides economic and demographic information for every existing and proposed fixed-guideway transit station in the United States.

The Database includes 4,416 existing stations and 1,583 proposed stations in 54 metropolitan areas, as of December 2011.

Data are available at three geographic levels: the transit zone (the half-mile or quarter-mile buffer around the individual station), the transit shed (the aggregate of transit zones), and finally, the transit region (aligns with the Metropolitan Statistical Area boundary).

Nearly 70,000 variables are derived from nationally available data sets including the 2000 and 2010 Decennial Census, the 2009 American Community Survey, the 2000 Census Transportation Planning Package, and the 2002–2009 Local Employment Dynamics.

Deteriorating or Improving?

14

Transport Sustainability Trends in Global Metropolitan Areas

JEFFREY KENWORTHY[1]

There is little doubt about the growing recognition worldwide that cities need to become more sustainable in their transport patterns. The world's seven billion people are increasingly locating in cities, thus placing unprecedented pressure on urban transport infrastructure and making it increasingly difficult for cities to help to ameliorate urgent global challenges such as climate change and the host of other more regional and local environmental issues facing people all over the planet. Arguably for pure economic reasons, we must also address the way cities develop, for cities are clearly the economic engines of nations.[2] If human economic well-being is to improve, it must do so in a way that is not utterly damaging to the ecological systems that underpin all life. Indeed, cities must start to regenerate and repair their environments and become what are now termed Regenerative Cities,[3] not just cease to do further damage.

The issue of urban mobility must be addressed, which requires a consideration of land-use issues in cities.[4] We know that the cities of North America, Australia, and many other places are already highly automobile-dependent, leading to massive problems of land and resource consumption and pollution, while a huge proportion of the world is undergoing rapid motorization (China, India, Brazil, etc.), leading to dramatic congestion and spiraling CO_2 production from passenger transportation.

This chapter examines energy use and transportation systems of cities in the currently wealthier or higher-income cities of the world, especially the United States, and how they are evolving today at this critical juncture in human development where peak oil, climate change, the global financial crisis (GFC), and many other problems threaten to challenge every assumption that has governed urban development since the Second World War. It looks at some of the critical changes that have been occurring between 1995–1996 and 2005–2006 in a sample of ten US cities, five in Canada, four in Australia, and twelve in Western Europe, as well as Singapore and Hong Kong for some very distinct contrasts. It does this by examining a huge array of data measuring land use, private and public transport service, resource usage, and infrastructure, as well as factors resulting from transport patterns such as energy use. Collectively, these data help us to characterize transport patterns in cities and to assess their land-use / transport-system sustainability. In this chapter we ask the question "Are these cities becoming more or less sustainable in land use and transport terms?" and answer it with an overall transport-sustainability report card on the 33 cities, based on the data analysis.

To pursue these objectives, the chapter presents results of a partial update of urban data contained in the Millennium Cities Database for Sustainable Transport by Kenworthy and Laube and an expansion of the number of US cities examined.[5] This original database contained 1995 or 1996 data (depending on the national census year of the city) for a wide range of land-use, transportation, economic and energy/externalities items for 100 cities around the world, including four cities in Australia (Brisbane, Melbourne, Perth, and Sydney), ten cities in the United States (Atlanta, Chicago, Denver, Houston, Los Angeles, New York, Phoenix, San Diego, San Francisco, and Washington), five Canadian cities (Calgary, Montreal, Ottawa, Toronto, and Vancouver), 32 cities in Western Europe, of which twelve are included in this paper (see table 14.1), and many cities in Asia, of which Hong Kong and Singapore are included here. (The term "cities" herein refers to whole metropolitan areas.)

Global Comparisons: Passenger Transport Energy Use and Urban Density

Of critical interest to the future of cities is the extent to which they are dependent on ever-more-expensive liquid fossil fuels. World oil production has either already peaked or will peak within the next few years. Oil prices hit almost $150 per barrel in June 2008 (Brent crude oil prices) and sent significant shockwaves throughout the global economy,[6] and were a contributory factor in the sub-prime mortgage meltdown in the United States,[7] which ricocheted through the financial sector across the globe and led to the Great Recession. As of February 2012, oil prices have again crept back to around $130 per barrel.

All cities in this study, with the exception of the Australian and European cities, have on average reduced their per-capita demand for energy in private passenger transport—US cities by 11 percent, Canadian by 5 percent, and Singapore/Hong Kong

Table 14.1 The 33 core cities contained in this update study

USA	Canada	Australia	Europe	Asia
Atlanta	Calgary	Brisbane	Berlin	Hong Kong
Chicago	Montreal	Melbourne	Frankfurt	Singapore
Denver	Ottawa	Perth	Hamburg	
Houston	Toronto	Sydney	Munich	
Los Angeles	Vancouver		Zurich	
New York			Copenhagen	
Phoenix			Helsinki	
San Diego			Oslo	
San Francisco			Stockholm	
Washington			London	
Manchester				
Stuttgart				

by 6 percent. A combination of improving technological efficiencies in motor vehicles and stabilizing car use (see fig. 14.1) appear to be behind these changes. In the case of the US and Canadian cities, their vehicle fleets were so grossly inefficient that trimming the "low-hanging fruit" from the consumption of fuel in these fleets in the 10 years to 2005 was a comparatively easy task, the main strategy of which was simply downsizing. By contrast, the Australian cities grew in car use by 14 percent and in energy use by 16 percent, which suggests that, on average, urban vehicular fuel-consumption rates per kilometer actually increased marginally. European cities increased their private passenger transport-energy use per capita by a more modest 4 percent.

Even with the decline in per-capita fuel use in cars in US cities, they are still by far the highest energy consumers in urban private-passenger transport, standing at over 53,000 MJ per capita in 2005, compared to the next highest, Australian cities, which average 36,000 MJ. Canadian cities are even lower at 31,000 MJ. European cities are a comparatively miserly 16,500 MJ, and Singapore/Hong Kong weigh in at a meager 6,000 MJ, nearly a ninefold difference compared to US cities. Such stark differences in the use of this critical nonrenewable resource point to very important lessons that must be learned from the low-energy consumers—lessons that relate to urban land use, the role of transit, and the role of walking and cycling—all lessons that are well within the capabilities of the largest energy consumers to learn and to implement.

Within the US cities, only San Diego and Washington actually increased their per-capita energy use (10 percent and 18 percent, respectively). In other cities the decline from very high levels was in some cases relatively dramatic, with Atlanta shaving off

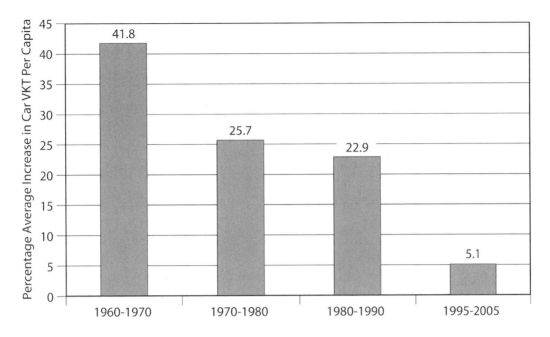

Figure 14.1   Car-use growth trends in a large sample of American, Canadian, Australian, and European cities, 1960–2005, using the Global Cities Database.

27 percent and Houston 34 percent from what were extraordinarily high levels of use in 1995.

As densities rise in cities, per-capita car use and transport energy decline.[8] It is thus important for sustainability that urban densities are on an upward trend, notwithstanding the fact that metropolitan scale densities tend to change slowly. The data show that urban population density is on the rise in Australian and US cities as well as in Singapore/Hong Kong, despite the latter's already very high densities. Canadian cities have more or less stabilized their densities, having on average declined only 1.5 percent in the ten years. Those in Europe declined a little more (3 percent) but also appear to be stabilizing in this regard. On the other hand, when one combines jobs with population to establish "activity density," we find it is rising in all the cities in this sample except for European cities, where the 1995 and 2005 average activity density for the 12 cities is identical. These so far modest trends of 1–8 percent increases in urban density and activity density are, however, significant because they represent a very important reversal of the history of these cities since the Second World War, when densities for the most part continuously declined under the influence of automobile-based planning and sprawling development. They also suggest that focused density increases within some cities are now being felt at a metropolitan scale in a reversal of previous downward trends.

Summary of Global Trends

Overall, how are cities performing in transport-sustainability terms over the period 1996–2006? Our study examines 25 important variables that have been used to develop a kind of "sustainability report card" on this sample of 33 global cities. It is relatively easy to judge whether the trends support or detract from greater urban transport sustainability. This report card is summarized in table 14.2, where each of the 25 variables are assessed on the basis of five different trend categories in relation to transport sustainability:

1. Consistently positive (CP)
2. Consistently negative (CN)
3. Genuinely mixed (GM)
4. Mixed but generally positive (MGP)
5. Mixed but generally negative (MGN)

We can say that in ten of the 25 variables, or 40 percent, the trends are in a consistently positive direction for sustainability for the groups of cities as a whole, with some natural within-group variability. Of these ten variables, eight are related to aspects of the transit system, such as growing service supply, growing use and so on. Increasing activity density and a reduction in transport deaths per 1,000 persons are the other two consistently positive stories to come out of the research.

Furthermore, another ten variables achieve a mostly positive trend toward sustainability, with generally only one group of cities diverging from that trend. Six of these mostly positive variables are related to transport infrastructure items, both private and public transport infrastructure. This means that of the variables examined, 20 out of the 25, or 80 percent, demonstrate a generally positive direction with respect to transport sustainability.

Interestingly, only two variables demonstrate a consistently negative trend—car usage in terms of vehicle-kilometers per capita and the proportion of jobs located in the central business district (CBD) of cities. Only one further variable exhibits a mostly negative trend and that is car ownership. In must be said on the positive side, however, that car use also had already begun to show some positive signs by 2005 in terms of the slowing of the growth rate and this appears to have continued in the post–2005 period.[9] The decline in the proportion of jobs in the CBD does not necessarily mean a

Table 14.2 A transport sustainability report card on 33 global cities, 1995/6 to 2005/6

Variable	Trend (1995/6–2005/6)	City group against trend
Urban form factors		
Urban density (persons per ha)	MGP	Canadian/European
Activity density (persons + jobs per ha)	CP	None
Proportion of jobs in CBD (%)	CN	None
Private transport infrastructure factors		
Length of road per person (meters)	MGP	Canadian
Length of freeway per person (meters)	MGP	Canadian
Parking spaces per 1,000 CBD jobs	MGP	European
Cars per 1,000 persons	MGN	Canadian
Public transport infrastructure factors		
Total length of reserved public transport route per person (m/1,000 persons)	MGP	Australian
Ratio of reserved public transport infrastructure to freeways	MGP	Australian
Private transport use factors		
Car vehicle-kilometers per person	CN	None
Car passenger-kilometers per person	GM[a]	US/Australian/European
Public transport service, use, and performance factors		
Total public transport vehicle-kilometers per person	CP	None
Total public transport seat-kilometers of service per person	CP	None
Total rail seat-kilometers per person	CP	None
Total bus seat-kilometers per person	GM	Canadian/Australian
Total public transport boardings per person	MGP	Singapore/HK
Total rail boardings per person	CP	None
Total bus boardings per person	MGP	Singapore/HK
Total public transport passenger-kilometers per person	CP	None
Total rail passenger-kilometers per person	CP	None
Total bus passenger-kilometers per person	CP	None
Proportion of total motorized passenger-kilometers on public transport (%)	CP	None
Ratio of public transport system speed to road traffic speed	MGP	Canadian
Energy and externality factors		
Total private passenger transport energy use per person (MJ)	MGP	Australian/European
Total transport deaths per 100,000 people	CP	None

Summary of trends	Number of cities	Percentage
Consistently positive (CP)	10	40%
Consistently negative (CN)	2	8%
Genuinely mixed (GM)	2	8%
Mixed but generally positive (MGP)	10	40%
Mixed but generally negative (MGN)	1	4%
Total	25	100%

[a] For the Genuinely Mixed category, the city groups listed against the trends are the ones that go **against a positive** sustainability trend.

decline in the absolute number of jobs located there, and if a good proportion of the non-CBD jobs are shifting to other significant sub-centers, the trend is not necessarily bad for sustainability. If they are scattering, however, it will lead to more car use. The two genuinely mixed trend variables are car usage, expressed as car passenger-kilometers per capita, where the increases are very small and two city groups show declines, and bus service provision in terms of seat-kilometers per capita.

In summary, it can probably be said that at least these 33 cities, and possibly many others, are at some sort of tipping point where there is evidence of some turnaround in factors that have long shown negative trends. Urban transport and land-use policy, as well as demonstration projects, will be critical in how these trends play out over the next ten years. The question is: Can the positive trends be maintained and strengthened and the negative trends extinguished, and might such trends also herald not just a positive picture for sustainability, but even the beginnings of regenerative cities, led by a decline in automobile dependence? Let's take a closer look at the United States.

Transit in US Cities, 1995–2010

This section first examines some 1995 to 2010 trends in transit for the ten American cities listed in table 14.1. It then considers a larger sample of US cities (20 in all) and examines the changes in the same three key transit system performance variables from 2005 to 2010 (the amount of service provided by mode per capita, the number of annual boardings per capita [trip rate], and the annual passenger-kilometers of travel per capita [travel distance]). It is now established that transit use as a whole in the United States has, at least recently, been increasing.[10] It is, however, interesting to see the perspective within major metropolitan areas of the country and in particular some regional differences.

Transit Performance in Ten Major Metropolitan Areas, 1995–2010

The first three graphs (figs. 14.2 to 14.4) provide average data for these three variables for the ten large US metropolitan regions listed in table 14.1 over a 15-year period.

As can be seen, total per-capita provision of transit has grown over the period by about 27 percent, though the 2005–2010 period experienced less growth than the previous ten years. Bus service has grown more than rail service, most likely because of servicing demands in new low-density development.

Figure 14.3 shows that total per-capita trips by transit have also risen steadily over the fifteen years (about 14 percent), though they seem to be tapering off up to 2010, perhaps due to a lower demand in the critical commuting market, which in turn is linked to higher unemployment in the United States due to the Great Recession. What

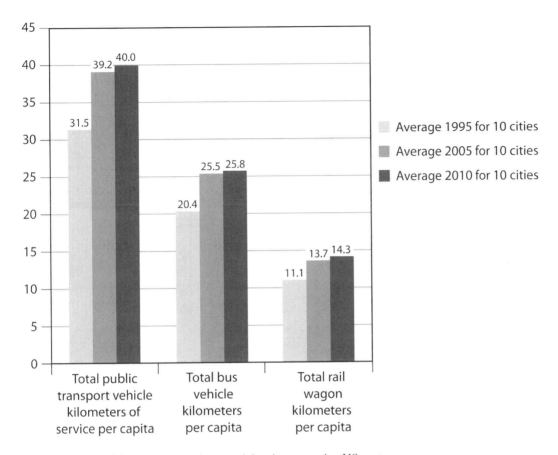

Figure 14.2   Public transit service provision in ten major US metro areas, 1995–2010.

seems to be very clear, however, is that bus usage has struggled (5 percent less in 2010 than in 1995), while rail usage has steadily increased (44 percent higher in 2010 than in 1995), perhaps due to the greater attraction and reliability of railroads running on dedicated rights-of-way.[11]

Figure 14.4 shows transit use in terms of how much mobility or travel distance people use it for. Again, the growth in mobility by transit in these major US metropolitan areas is clear. While buses have grown a little in 2010 compared to 1995 (3 percent), bus use seems to be at a plateau, having declined a little from 2005 to 2010. On the other hand, despite whatever economic down-pull may be at work in the United States, rail has continued to grow steadily (39 percent over the 15 years). This may indicate travelers' pent-up demand for a quality transit alternative in the United States, which is finally being realized in the many new urban rail systems that have been built in US cities in recent decades.

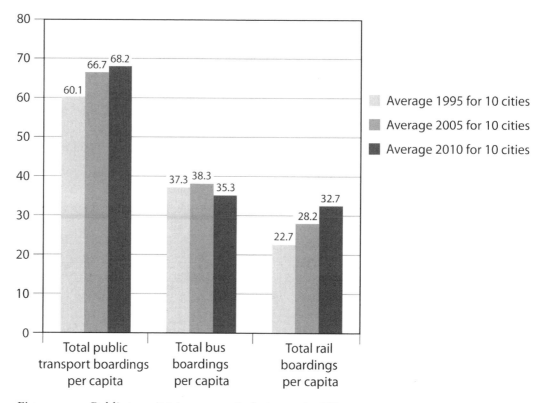

Figure 14.3   Public transit trips per capita in ten major US metro areas, 1995–2010.

It is also interesting to consider the usage factors in relation to service. For an increase of 3.2 wagon-kilometers per person from 1995 to 2010, rail picked up ten more boardings per capita. Bus service, on the other hand, rose by 5.4 vehicle-kilometers per capita and the systems *lost* two boardings per capita. Perhaps there is a fundamental issue here. Buses are attempting to bring service to the more dispersed parts of US metro areas and need to run a lot more service for a diminishing return (refer to previous data on the increase in US cities' bus vehicle-kilometers of service). On the other hand, rail is concentrating at least some land use and is bringing increasing numbers of residents and businesses within its reach through TOD, thereby getting a much better transit-use return on "investment."

Transit Performance in Twenty US Metropolitan Areas, 2005–2010
It is interesting to expand this analysis by examining more cities and focusing on the more recent five-year change from 2005 to 2010. This has been achieved by adding the metropolitan areas of Portland, Seattle, Detroit, Pittsburgh, and Honolulu, as well as

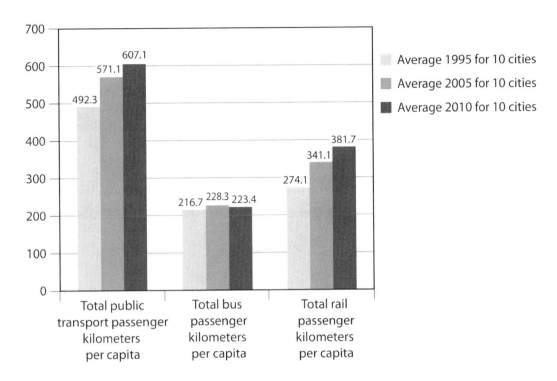

Figure 14.4   Public transit passenger-kilometers per capita in ten major US metro areas, 1995–2010.

an additional five US cities in the southern states of considerably varying size (New Orleans, Charlotte, Columbia, Memphis, and Miami). These southern cities join Atlanta and Houston, which are already in the ten cities in table 14.1.

Figure 14.5 shows the service levels provided per capita in this whole expanded sample of 20 US cities in 2005 and 2010. Here we see a somewhat different perspective with regard to overall per-capita service provision, which has on average declined marginally in the 20 cities due to the decline in the provision of bus service. However, the provision of rail did grow marginally over the five years, thus moderating the overall decline in the provision of transit service.

When we consider the usage of these systems we again find that there was, on average, a small decline in the five years in per-capita transit boardings (fig. 14.6), and that the average usage of the systems for these 20 cities is considerably lower than that of the original 10 cities, due mainly to the addition of the weak transit systems in southern US cities. Buses in particular declined in usage, but again, rail systems were resilient and recorded a 16 percent increase in boardings per capita over the five years

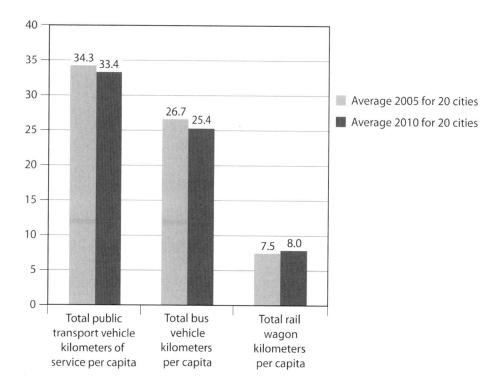

Figure 14.5 Public transit service-kilometers per capita in 20 major US metro areas, 2005–2010.

(while rail service per capita rose only 7 percent). A slightly different pattern emerges when the travel per capita by transit is considered, as seen in figure 14.7. In this case, total transit travel per capita has risen a little, while buses have remained almost stable. This indicates that while actual boardings per person have fallen, the distances people are traveling seem to have increased. Rail maintains its growth trajectory, as in all the previous analyses.

To finish this section, figure 14.8 shows the distance traveled per capita by transit in the southern US cities only and reveals a significant decline both in the total figure and particularly in the buses. Rail again registered a small increase even in these transit-embattled southern cities, though the average per-capita rail travel in the ten US cities in figure 14.4 is fully eight times higher than in the seven southern cities of Atlanta, Houston, New Orleans, Charlotte, Columbia (SC), Memphis, and Miami.

It is interesting to note that New York was by far the best-performing US metro area in transit usage increase from 2005 to 2010. Annual transit boardings rose from 168 to 192 per capita, up still further from 131 in 1995. On the other hand, Columbia

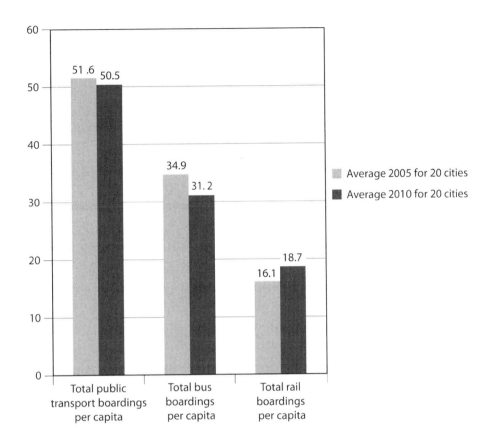

Figure 14.6   Public transit boardings per capita in 20 major US metro areas, 2005–2010.

recorded in 2005 a staggering annual total of 5 transit trips per capita, which actually *declined* to 3 trips per capita in 2010, yielding what is tantamount to a US city of 650,000 people without any transit use to speak of!

Policy Implications

The data assembled for this paper and presented in the previous sections point to a number of policy implications, which are drawn out here with a particular focus on the US cities.

Urban Form, Centers, and Parking

It is clear that urban population and job densities in this significant sample of cities are showing an upward trend in most cities, though the upward trend is small on a metropolitan scale. For more-sustainable transport patterns and many other sound

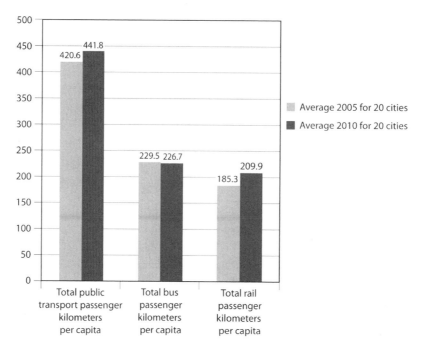

Figure 14.7   Public transit passenger-kilometers per capita in 20 major US metro areas, 2005–2010.

ecological reasons such as reduced land consumption, protection of food-growing areas, and a more livable public realm, there is a need to continue to grow densities, not in an unplanned or uniform way across the urban landscape, but more particularly in focused centers in order to create a more transit-oriented urban form or polycentrism. This should lead to a greater modal share for non-auto modes and also to shorter trip distances.[12]

Related to this question of density increases, the sub-centers and CBDs of cities should not be overlooked as a location for work and an increasingly important site for population growth so that centers also become better neighborhoods. There should be limits on ad hoc dispersal of work into locations that are really only conveniently and competitively accessible by car. The preeminence of the central city is declining in relative terms (even where there is still a net growth in actual jobs), and job growth outside the CBD needs to be captured as much as possible in viable centers served by quality transit (rail modes are generally the best), with high-quality public realms and feeder facilities (both rail and bus) to promote greater transit and nonmotorized modal share for trips to and within each center.

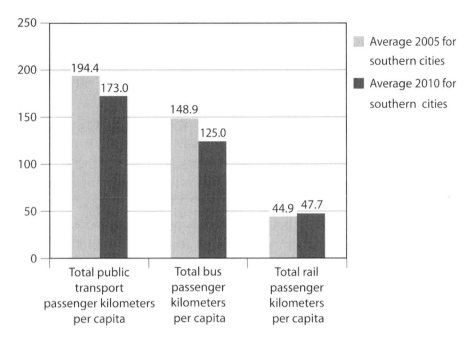

Figure 14.8   Public transit passenger-kilometers per capita in seven southern US metro areas, 2005–2010.

For US cities, these policy implications are critical—of all the cities in the sample, US cities have dispersed the most away from their CBDs and the resulting scattered patterns of employment location *may* have led in some cases to shortened commute distances and times, but they have radically increased dependence on the car per se, which negates any transport benefits to society.

Further related to these policy implications, it can be said that parking supply in the CBDs of cities globally is, on average, generally on a downward trend, which is a good trajectory for sustainability on many levels, especially better transit use to the center. Where surface parking is replaced by public facilities, housing, and other uses, the whole livability and human attractiveness of the city also rises. For example, Portland in the United States has seen a declining and car-oriented city center change into a vital, colorful, and economically successful hub over the space of 30–35 years. The creation of Pioneer Courthouse Square from a parking lot, the construction of the River Place development along the Willamette River, the tearing down of the Harbor Freeway in front of downtown, the installation of both an LRT and a streetcar system in the central and inner city as well as suburban areas, the Pearl District revitalization around the streetcar, the ringing of remaining CBD surface parking lots with semi-permanent food caravans, and generally splendid urban design and landscaping throughout, has secured a world-class CBD for Portland.[13]

However, the majority of the US cities in table 14.1 actually increased in CBD parking supply, a trend that clearly calls for some policy attention for all the reasons just discussed.

The general global decline in CBD parking provision relative to jobs, as suggested by the data in this chapter, certainly also strengthens the possibility of residential and job growth, since parking can occupy vast amounts of land in CBDs. The implication is that the CBDs of cities need to continue growing in jobs but not in parking supply, and surface commuter parking areas need to be built out with new housing to help the trend and to make the CBD a better neighborhood and not just a business center.[14] Where appropriate it may also become feasible to demolish CBD parking structures to make way for more employment space, better public environments, or housing complexes. For the US cities, there is abundant scope to realize such a policy direction.

Private Transport Infrastructure
Car ownership in cities is still clearly on the rise, except for the Canadian cities where there was a small decline. This has many implications, including the potential for more car use as well as extra land consumption across metropolitan areas, because for each new car, more car parks are needed (each car has on average multiple parking places located across the city at shopping areas, workplaces, hospitals, etc.). The economic incentives and disincentives for cars need to be closely examined. Free parking at workplaces should also be examined.[15] Generally speaking, cities should look at the various successful schemes worldwide for limiting the growth in car ownership such as those in Singapore and Shanghai, which seek to match new car registrations with the attrition rate of old cars through an auctioning or lottery system for the right even to buy a car. US cities need to see whether adoption or adaptation of such schemes would be feasible. Also, the potential of car-sharing systems should be more widely promoted and supported by government because they have already resulted in falling car ownership in cities such as Bremen in Germany.[16]

Related to limiting car ownership and parking, overall road provision in cities per person needs to continue its general stabilization/decline by a combination of minimizing new arterial and highway construction and reducing local road provision through more compact land use. In particular, the construction of new freeways is still a major issue in some cities, with substantial increases in this factor in some US cities over the ten years (e.g., Denver and Washington). Freeway construction is preventing cities from making more-substantial shifts toward sustainability in transport.

One could say that US cities in particular need *biased* investment in new transit systems, especially rail, along with a cessation in the self-defeating construction of more freeways, which undermines the success of new transit systems and prevents the full potential land-use advantages of rail systems from occurring. Vancouver is a leader in TOD today, at least partly because it built no freeways and it anchors a huge amount of new development around an ever-expanding rail system that provides the fastest travel option in the region for many trips.[17]

Transit Infrastructure, Service, and Use

On the reverse side, apart from the Australian cities, most cities are clearly doing quite well in the construction of new, dedicated rights-of-way for transit, and in the cities of the United States, Canada, and Western Europe, as well as in Singapore and Hong Kong, there is an improving ratio between the provision of high-quality transit and that of freeway provision. In order for this to continue, not only do new transit systems need to be constructed, especially rail systems, but policies in each city need to be considered to determine how they might actually remove strategically located pieces of freeways, as was done in Seoul[18] and has been done in critical areas of many other cities around the world (e.g., Portland, San Francisco, Milwaukee). Such removals not only assist the competitive position of non-auto modes of transport, but in general they radically improve the urban design and livability qualities in the areas affected by these removals, without any negative effects on traffic flow and sometimes even increases in average traffic speed in the city (as in Seoul). The evidence suggests that traffic behaves more like a gas than a liquid when its space is restricted.[19]

The waterfront areas in downtown San Francisco and Portland have been greatly enhanced by removal of the elevated Embarcadero and Harbor Freeways, respectively, and this has opened up new possibilities for people and businesses in those areas and their relationship with attractive water environments. Portland's Tom McCall Park, the site of the annual Rose Festival, is a great addition to the city where the Harbor Freeway once blighted the waterfront. Every US city has its opportunities for such removals, but the politics is often bloody and torturous.[20]

In terms of the amount of service that is supplied by urban transit systems, the picture is a rather positive one, with all groups of cities expanding their service between 1995 and 2005, particularly the amount of rail transit service offered to their populations. All transit systems need to move increasingly toward non-timetable frequencies of 10 minutes or less, and the connections and integration between transit systems need to be improved. Further, this integration needs to be approached on both a physical level in terms of people actually getting between the services conveniently and comfortably, with maximized opportunities for linking trips (e.g., shopping, business meetings, personal visits, etc.), as well as better fare integration, which provides transit mobility at a price that is very attractive compared to the car and which rewards committed users with big discounts. Successful transit systems everywhere in the world are built on transfers between different modes and operators, and on building up a substantial cohort of committed users, generally with prepaid tickets. Physical connections between transit modes are best achieved in centers that operate on a time-pulse transfer system (where all transit services depart simultaneously at a regular time interval such as 5, 10, 15, 25, 35, 45, and 55 minutes past the hour), or that have saturation levels of service that make timetables irrelevant, that is, better than 10-minute frequencies.

The critical thing is that to grow transit use, it is not enough to merely increase service (sometimes increasing per capita levels of transit service can be a reflection of more scattered land-use patterns, which require far greater distances of travel by transit vehicles to pick up greatly dispersed potential passengers). The quality of that service is important and depends on how it relates to the urban structure of the city, as is seen in the Canadian cities data where transit usage is high relative to infrastructure and service provision.

Transit use in cities has, on average, headed in a positive direction in all groups of cities in this sample, with the provisos regarding the US cities already discussed. In particular, the rail modes have improved universally in both their trip-making rate as well as in the distances traveled by users (passenger-kms), while buses have fared less well. While transit has universally gained a little over the car in its share of total motorized passenger-kilometers, the actual increases are modest. Policies to increase the quality and offer of transit clearly need to be supported everywhere with complementary land-use policies to better integrate development, especially in dense centers throughout urban regions, revised parking policies, a cessation of freeway building, and an effort to charge more realistic prices for the true cost of urban motoring at least in peak periods, all of which are strongly related to each other as policy levers and all of which need to be pulling in the same direction.

The more detailed picture for transit within the United States, based on an expanded sample of cities, reveals very clearly that:

- Rail systems everywhere in the country are the backbone of improvements in transit service and growing usage. Rail generally has been shown to be the best-performing mode for regenerating transit systems,[21] and both global and US data here are highly supportive of this.

- The already more transit-oriented cities such as New York, Chicago, and San Francisco continue to improve in transit relatively more than the less transit-oriented cities in the United States, leading to a conclusion that success tends to build on success and decline is hard to reverse. Los Angeles offers evidence, however, that such decline can be reversed, since it showed by far the biggest percentage increase in boardings per capita from 1995 to 2005 (39 percent) of all the US cities examined here.

- Buses struggle to record greater per-capita usage nearly everywhere and appear to need the support of good rail systems to find their important central role in transit systems, primarily as excellent feeder systems.

- Southern US cities have extraordinarily poor transit service and usage, with a glaring lack of rail systems. In an already transit-poor nation generally, they still tend to be only half as well performing in transit as the "average" US city, a very sad achievement. Policy interventions are needed to provide better transit options for this populous region of the United States that has, in particular, significant socioeconomic inequalities.

Private Transport Use

Measures are needed to curb the still-growing (but slowing) car use per person in most cities. Car use can be diminished with the planning-oriented policies already outlined, but there is also much that can be done on an economic level, however unpopular politically. This includes charging the "right prices" for car use through congestion charges as in London's CBD, instituting road-pricing schemes, regulating the price of fuel, and utilizing other "carrots and sticks" such as the Certificate of Entitlements for car purchase in force in Shanghai and Singapore and now also in Beijing. Indeed, based on the many positive trends revealed in this study of cities, it could be argued that there should have been more strong increases in transit use and bigger reductions in car use. It could be postulated that "road pricing" is the missing link in just about all cities in order to reinforce and make the most of positive achievements in other areas to limit car use. This is especially true of cities in the United States, which still enjoy by far the cheapest fuel prices in the developed world,[22] and generally very low costs of driving overall.[23]

Energy and Externalities

Australian cities and, to a much lesser extent, European cities, stand out as the only cities in the sample where private-passenger transport-energy use is still growing. In all the other groups of cities, per-capita private-passenger transport-energy use has declined. Technological gains in the fuel efficiency of cars are capable of delivering reductions in energy use, as witnessed for example in the US cities. Here, the car fleet in 1995 was so profligate in its use of fuel[24] that, when combined with an almost stable car-use picture, energy use per capita declined by a significant 11 percent as the "low-hanging fruit" of too many oversized, overpowered vehicles began to be picked off (there was enormous scope for improvement!). The same can be said for the Canadian car fleet, and to a lesser degree the Australian car fleet.

In general terms, the evidence is that technological changes that produce more fuel-efficient cars can easily be eaten up and exceeded by increases in car use.[25] Policies to limit car use and concurrently provide better propulsion systems and other energy-efficient design changes in cars are needed in order to really tackle the transport energy problem. Going further, policies and technologies that promote the use of renewable energy in transport need to be pursued, but such technologies should not be seen as a panacea to the energy problem in transport.

For example, the current strong push for "electro-mobility" in Europe, especially in Germany, is implicitly cast by engineers and technologists as a problem of replacing each kilometer driven by the current fleet of internal-combustion-engine vehicles with the same number of kilometers driven by electric vehicles. In practice, electric cars only make sense in a scenario of significantly scaled-down demand for car-based private motorized mobility, even if those cars are more "environmentally friendly." Substituting the current massive demand for private mobility based on liquid fossil fuels is not possible nor desirable for many reasons (e.g., space demands), but it may be possible to meet the energy demands of much-reduced car use per se.[26] This conceptualization of the problem is critically important for US cities, which have massively higher car use than any other cities in the world.

Notes

1. The author wishes to acknowledge the detailed work of my research assistant, Ms. Monika Brunetti, in extracting all the US transit data used in this paper, and making major contributions to the data work on the other cities. Any errors are the responsibility of the author.

2. See, for example: Jane Jacobs, *Cities and the Wealth of Nations: Principles of Economic Life* (Toronto: Random House, 1984); Karlson Charles Hargroves and Michael H. Smith, *The Natural Advantage of Nations: Business Opportunities, Innovation and Governance in the 21st Century* (London: Earthscan Publications, 2005).

3. World Future Council, "Regenerative Cities" (fact sheet on Cities and Climate Change, World Future Council, Hafen City University, Hamburg, Germany, November 2010).

4. Jeffrey Kenworthy, "The Eco-City: Ten Key Transport and Planning Dimensions for Sustainable City Development," *Environment and Urbanization* Special Issue (April 2006): 67–85.

5. Jeffrey Kenworthy and Felix Laube *The Millennium Cities Database for Sustainable Transport* (CD ROM database, International Union of Public Transport [UITP], Brussels, and Institute for Sustainability and Technology Policy [ISTP], Perth, Australia, 2001).

6. See: oil-price.net/dashboard.php?lang=en, accessed June 10, 2012.

7. Peter Newman and Jeffrey Kenworthy, "'Peak Car Use': Understanding the Demise of Automobile Dependence," *World Transport Policy and Practice* 17, no. 2 (2011): 31–42.

8. Peter William Geoffrey Newman and Jeffrey Raymond Kenworthy, *Cities and Automobile Dependence: An International Sourcebook* (Aldershot, UK: Gower, 1989); Peter William Geoffrey Newman and Jeffrey Raymond Kenworthy, *Sustainability and Cities: Overcoming Automobile Dependence* (Washington, DC: Island Press, 1999).

9. Newman and Kenworthy, "'Peak Car Use'"; R. Puentes and A. Tomer, *The Road Less Traveled: An Analysis of Vehicle Miles Traveled Trends in the US* (report, Metropolitan Infrastructure Initiatives Series, Brookings Institution, Washington, DC, 2009); A. Millard-Ball and Lee Schipper, "Are We Reaching Peak Travel? Trends in Passenger Transport in Eight Industrialized Countries," *Transport Reviews* (November 18, 2010): 1–22.

10. See, for example: www.planning.org/news/daily/story.htm?story_id=170187495.

11. Jeffrey Kenworthy, "An International Review of the Significance of Rail in Developing More Sustainable Urban Transport Systems in Higher-Income Cities," *World Transport Policy and Practice* 14, no. 2 (2008): 21–37.

12. Peter Newman and Jeffrey Kenworthy, "Urban Design to Reduce Automobile Dependence," *Opolis* 2, no. 1 (2006): 35–52.

13. Preston Schiller, Eric Bruun, and Jeffrey Kenworthy, *An Introduction to Sustainable Transportation: Policy, Planning and Implementation* (London: Earthscan, 2010).

14. Peter William Geoffrey Newman and Jeffrey Raymond Kenworthy, "Parking and City Centre Vitality" (paper presented at International Parking Conference, Perth, Australia, October 1988).

15. Donald Shoup, *The High Cost of Free Parking*, updated ed. (Washington, DC: American Planning Association [Planners Press], 2011).

16. Personal communication, Michael Glotz-Richter, March 2012.

17. Schiller et al., *An Introduction to Sustainable Transportation*.

18. Ibid.

19. Jeffrey Kenworthy, "Don't Shoot Me, I'm Only the Transport Planner (Apologies to Sir Elton John)," *World Transport Policy and Practice* 18, no. 4 (2012): 6–26.

20. Jeffrey Kenworthy and Craig Townsend, "Montreal's Dualistic Transport Character: Why Montreal Needs Upgraded Transit and Not More High Capacity Roads," in Pierre Gauthier, Jochen Jaeger, and Jason Prince, eds., *Montreal at the Crossroads: Superhighways, the Turcot and the Environment*, chap. 1 (Montreal: Black Rose Books, 2009), 29–35.

21. L. Henry and Todd Litman, *Evaluating New Start Transit Program: Comparing Rail and Bus* (publication of the Victoria Transport Policy Institute, Victoria, BC, 2011), www.vtpi.org /bus_rail.pdf, accessed June 3, 2012; Jeffrey Kenworthy, "An International Review of the

Significance of Rail in Developing More Sustainable Urban Transport Systems in Higher In-come Cities," *World Transport Policy and Practice* 14, no. 2 (2008): 21–37.

22. See: www.gtz.de/fuelprices.

23. Kenworthy and Laube, *Millennium Cities Database*.

24. Ibid.

25. Peter William Geoffrey Newman and Jeffrey Raymond Kenworthy, "The Transport En-ergy Trade-off: Fuel-Efficient Traffic versus Fuel-Efficient Cities," *Transportation Research* 22A, no. 3 (1988): 163–74.

26. Jeffrey Kenworthy, *International Benchmarking and Best Practice in Adapting to a Future of Electric Mobility in Germany: Sustainable Transport or Just Electric Cars?* (consulting project to Land Hessen through the University of Applied Sciences, Frankfurt am Main, Hessen, Ger-many, 2011).

Policy Implications of the Nonmotorized Transportation Pilot Program

15

Redefining the Transportation Solution

Billy Fields and Tony Hull

A growing number of communities around the United States are seeking to increase the rates of active transportation (walking and cycling) in order to help address oil dependence and provide the co-benefits of improved livability, decreased pollution, and enhanced public-health outcomes. While rates of walking and cycling appear to be increasing across the country,[1] precise, causal data on the relationship between environmental interventions designed to spur active transportation use and walking and cycling rates are still being established.[2]

At a policy level, the need to test the impact of active transportation investments is at the core of the federally sponsored Nonmotorized Transportation Pilot Program (NTPP). The US Congress included funding for the NTPP in the 2005 Safe Accountable Flexible Efficient Transportation Equity Act: A Legacy for Users (SAFETEA-LU). The purpose of the NTPP is "to demonstrate the extent to which bicycling and walking can carry a significant part of the transportation load, and represent a major portion of the transportation solution, within selected communities." The program allocated approximately $100 million for the initial four years of the program to be split equally between four communities around the country. These communities (Minneapolis, Minnesota; Sheboygan, Wisconsin; Columbia, Missouri; and Marin County, California) were selected to represent a diverse set of urban, rural, city, and county geographies across the nation.

While the NTPP provides necessary resources to begin to expand and improve the active transportation system, many of the underlying factors that influence transportation-mode choice have been left fundamentally unchanged by the Pilot Program. Key policies and bureaucratic regulations that promote the auto-centered transportation system, such as continued subsidies for low-density land use and expanded highway transportation systems,[3] have continued throughout the course of the Pilot Program.

This complex mix of transportation policies is clearly evident in Minneapolis. Despite a progressive reputation and Minneapolis's designation as "America's Best Bicycling City" by *Bicycling* magazine, underlying policy choices and bureaucratic resistance to change continue to promote energy-inefficient transportation and land-use policies. The impact of this for Minneapolis is underscored by the Surface Transportation Policy Partnership analysis of land-use and transportation policy in Minneapolis. Authors point to "an embedded regime of government regulations and subsidies—tax laws, zoning ordinances, building codes, street design standards, pricing structures— all fostering maximum energy use and minimum efficiency."[4]

The interactions between the continuing status quo funding, agency practices that emphasize the auto-centered transportation and land-use system, and the intervention of a small active transportation funding stream result in a complex tangle of causes and effects that makes judging the policy impacts of active transportation investment to reduce oil dependence extraordinarily difficult.[5] Given the complexity of the interactions, what is the appropriate burden of proof for establishing that the program, as the congressional language suggests, "represent(s) a major portion of the transportation solution" in Minneapolis?

Minimizing Oil Use through Active Transportation: Measurement and Political Challenges

To address the question of assessing the Pilot Program in Minneapolis, this chapter explores the challenges and opportunities for defining success of active transportation investment in reducing oil dependence through a case study of the Minneapolis Pilot community. Two key challenges are explored.

First, identifying success requires the establishment of measurement indicators that trace the specific impacts of the Pilot Program on active-transportation usage rates. To evaluate the impact of these policies, researchers can attempt to measure change at a community level through surveys and/or attempt to measure more localized impacts through physical counts of users taken adjacent to new infrastructure. Both of these approaches were undertaken by the Minneapolis Pilot. This chapter reports on findings that show that the Pilot Program in all four communities increased walking and bicycling between 32.3 and 37.8 million miles between 2007 and 2010 and saved approximately 1.67 million gallons of gasoline.[6] Implications and context of these findings are discussed.

While quantitative measurement provides an important framework for understanding the impact of active transportation investment, systematic institutional practices within the departments of transportation, public works, and planning can either hinder or accelerate movement toward a more robust active transportation system. The default position within transportation and planning departments has been based around producing single-use, low-density land uses connected through high-volume automotive corridors. These features make auto-oriented transportation and sprawling land use not "so much a deliberate choice as . . . a product of bureaucratic inertia."[7]

The second section examines the challenge of altering the "bureaucratic inertia" of administrative practices and procedures within departments of public works that act as significant barriers to promoting a wider active-transportation infrastructure base. We evaluate the impact of the Pilot on changing these underlying practices through a series of 28 in-depth interviews with key Pilot stakeholders. Before we examine the specifics of these two areas, a short literature review section is presented to provide context to these issues.

Evaluating Active Transportation Investment: Measuring Policy Impacts

Over the last decade, research into the connection between the built environment and active-transportation use has grown dramatically. Spurred on by the public health community's need to better understand the obesity epidemic[8] and from the planning and transportation sectors' emphasis on understanding the impact of land use and transportation choice,[9] research has centered on determining the potential environmental determinants of active transportation use.

In general, a correlation between design characteristics of the built environment such as sidewalks, trails, and bicycle facilities, and rates of walking and bicycling has been found in the scientific transportation literature. At this point, studies have not established clear causation between built-environment elements and active-transportation use. Other factors such as self-selection in neighborhood choice are still being examined. Despite this lack of scientific certainty, the accumulated evidence suggests a correlation between the built environment and active transportation without clear causation.[10]

Despite the lack of conclusive causative evidence, a number of studies have begun to build a strong conceptual framework for understanding how the built environment impacts transportation choice.[11] One way to help understand the complex factors influencing active transportation use is the "Three Ds" framework proposed by Cervero and Kockelman.[12] This influential framework suggests that mode choice is impacted by the density of land uses, diversity of destinations, and design of a community. Frank and Kavage propose adding a fourth "D" to the framework.[13] They suggest that proximity to destinations may also have an impact on active-transportation choice. In an environment where land uses are moderately dense, a wide variety of potential destinations are fairly close, and roadway and streetscapes are designed to make walking and bicycling attractive and encouraging, an individual will experience an increase in potential transportation-mode choices as compared with a less dense, more automotive landscape.

The "Four D" construction of density, diversity, design, and proximity of destinations provides a useful conceptual framework for understanding how the built environment impacts travel choice. While the above-mentioned studies show the difficult but manageable task of measuring a "Four D" built environment, the key metrics used in many of these studies differ from traditional transportation indicators. Traditional transportation measurement centers on mobility-based variables that focus on overall distance traveled. In this paradigm, "successful modes" move large volumes over long distances.

A more nuanced approach to defining transportation success focuses on evaluating the extent to which the transportation system provides access for people to goods, services, and amenities. This accessibility framework focuses on reducing overall travel by expanding proximate destinations and modal choice.

Kooshian and Winkelman argue that this more access-based approach to city building is central to both understanding the potential energy saving of synergistic built environment and travel-policy changes as well as potential economic benefits associated with denser development.[14] They argue:

> Mobility often is held out as the object of transportation policy, but mere movement in and of itself does not equate to economic productivity, while accessibility is critical to it. . . . Accessibility is highest when more homes are closer to shops and offices (*mixing land uses*), when there are multiple ways to get around (*a variety of transportation choices*), and the distances to be covered to accomplish daily life are kept to a minimum (*compact design*). Under these circumstances, people have greater access to economic activity and recreation, but with lower VMT.[15]

The accessibility transportation solution provides an avenue for decreased oil dependence and increased economic productivity. Ewing et al., for instance, have found that residents of smart growth–oriented communities drive between 20 and

40 percent less than residents in more sprawling landscapes.[16] While creating and co-ordinating land-use and transportation policies to support increased accessibility is a large challenge within the current policy climate, increased long-term demand for these types of connected, accessible, "walkable urban" places[17] will present opportunities in the years ahead.[18]

From an engineering perspective, the change to an accessibility framework is not technically complicated. Former New York City traffic commissioner Sam Schwartz points out that "engineers have the technical know-how to implement active transportation. The methods are deceptively easy: Build good transit systems and integrate them into existing infrastructure. Design transportation systems with pedestrians and cyclists in mind. Construct multiple, direct connections within dense mixed-land-use developments. Coordinate transit, walking, cycling, and automobile networks.[19]

There are currently several promising policies that are pointing the way to this new paradigm of development. The federal TIGER program and Sustainable Community Grants provide a strong foundation for beginning to understand how to create these types of cross-cutting, interagency working relationships necessary to build more-accessible places. The Nonmotorized Transportation Pilot Program, as a precursor to these current policy initiatives, highlights both the promise of promoting increased accessibility and the challenges of working to institutionalize accessibility as central goal within the transportation sector.

Impact of the NTPP: The Minneapolis Experience

The Pilot program represents an important opportunity to examine the impacts of a community-wide effort to change the built environment and test the impact of those changes. Evaluating the impacts of this program poses challenges in terms of both defining appropriate indicators and in the actual measurement of those indicators. This section provides an overview of the program, a description of the quantitative analysis tools used to evaluate the program, and key implications of the analysis approach for judging the success of the program.

Bike/Walk Twin Cities: Overview of Structure of the Program

The Minneapolis Pilot, known locally as Bike/Walk Twin Cities (BWTC), is administered by Transit for Livable Communities (TLC), a nonprofit organization in Saint Paul. BWTC is governed by a 12-member board of directors, including the City of Minneapolis and 13 municipalities in three adjoining counties. Program eligibility is focused on access to and from Minneapolis.

The selection of TLC to administer the Minneapolis program was intended to demonstrate the capacity of nongovernmental organizations to bring a nontraditional approach to innovative implementation of a federal transportation program. TLC, with a strong presence in the region, was perceived as both a credible and a geographically neutral choice.

The nontraditional approach to administration, combined with the multi-jurisdictional program area, provided for unique opportunities and numerous challenges. The result was a complex arrangement of process and partnerships framed around the program. This complex arrangement structure, while difficult to work through, proved essential to crafting innovative solutions and facilitating delivery of more access-based projects and programs that had not been funded within the traditional highway funding approach.

To facilitate TLC's role as program administrator, the City of Minneapolis was designated to act as a fiscal agent, as law prohibits federal transportation dollars from being allocated directly to non–tax authority entities. Contracts between TLC and the City of Minneapolis, as well as between the City of Minneapolis and MnDOT, were required to enable the flow of federal funds. Additionally, because TLC was new to the federal project process, an agency partner group was formed that included staff from TLC, City of Minneapolis, MnDOT, FHWA, and Metropolitan Council. The group met monthly to monitor progress of program implementation and develop innovative strategies to fund nontraditional projects under the rigid federal transportation-funding guidelines.

In 2006 the TLC Board of Directors appointed a project advisory committee, known as the Bike-Walk Advisory Committee (BWAC), to provide expertise and stakeholder input from relevant disciplines and interests. The BWAC served as advisors to the board regarding funding strategy and process for project selection, assisted in reviewing project applications, and provided funding recommendations to the TLC board. The BWAC, comprised of planners and engineers from city, county, regional, and state agencies; transit representatives; pedestrian and bicycle advocates; various nongovernmental stakeholders; and elected officials conducted business in open public meetings similar to advisory bodies of Metropolitan Planning Organizations.

Prior to the arrival of the NTPP program, travel by bicycling and walking had been viewed more as a local issue than a regional issue, and Metropolitan Council policy did not allow for federal transportation funds to be used for local bicycle- and pedestrian-planning purposes. As a result, there has not been a regional bicycle or pedestrian master plan for the Twin Cities.[20] The City of Minneapolis had, however, made strong investments in bicycling infrastructure, including a well-developed trail system and initial components of an on-street network. The on-street investments had typically been undertaken on a project-by-project basis, rather than as part of a comprehensive plan for bicycle and pedestrian facilities.

BWTC: A Change in Priorities

BWTC ushered in a new era in project planning. Instead of the on-street projects emerging in an ad hoc fashion, BWTC sought to establish program goals that would guide the distribution of resources for projects. All projects were guided by the following three overarching program goals:

1. Maximizing existing roadway for all users: creating an interconnected, multimodal network to facilitate accessible, year-round short-trip options.
2. Creating regional legacy though planning, performance measures, and innovation.
3. Building local and regional capacity within the transportation professional community, political leadership, and the general public.

The result of this new approach to program planning led to significant gains in active-transportation infrastructure. Prior to the program in 2005, Minneapolis had 46 lane-miles of on-street facilities and 75 miles of off-street trails. As of 2011, these numbers had significantly increased to 130 lane miles of on-street faculties and 86 miles of off-street trails (table 15.1). Figure 15.1 shows the increased connectivity created by the enhanced system.

While BWTC helped to begin to address active transportation in a systematic way, the infusion of federal funding in the Twin Cities for walking and bicycling was relatively small as percentage of overall transportation funding for the region. Analysis of federal funding data from the *Fiscal Management Information System (FMIS)* provided by the Federal Highway Administration (FHWA) shows that approximately $1.6 billion in overall transportation funding was obligated to the Twin Cities with a total of approximately $66 million dedicated to bicycle and pedestrian projects between August 2005 and September 2011. Put another way, walking and bicycling projects accounted

Table 15.1 Minneapolis bicycle facility level change

Type of facility	Pre- BWTC (miles)	Fall 2011 (miles)	Increase	% BWTC Funded
Off-street bicycle facilities	75.4	86.4	15%	1%
Lane-miles on-street bicycle facilities	46.1	129.5	181%	72%
Total mileage	121.5	215.9	78%	64%

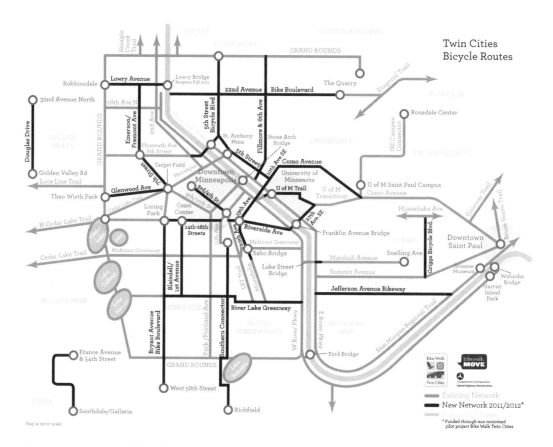

Figure 15.1   Twin Cities bicycle routes.

for only 4 percent of all federal transportation funding during the height of the Pilot Program in the Twin Cities.

While BWTC was the highest profile active transportation program, BWTC accounted for only about 30 percent of federal bicycle/pedestrian funding during the period. Of the $66 million obligated for bicycle/pedestrian projects, approximately $21 million was dedicated to BWTC projects. The remainder ($44.8 million) was obligated to non-Pilot active-transportation funding categories such as transportation enhancements, safe routes to school, and ARRA (stimulus).

It should be noted that the percentage of federal funds going to bicycle/pedestrian categories in the Twin Cities even without the Pilot is about double the nationwide percentage of federal bicycle/pedestrian funding. The Twin Cities without the Pilot is spending about 3 percent of federal funds on bicycle/pedestrian projects, while the overall national percentage is about 1.5 percent.

Evaluating Success: Pilot Data Analysis and Modeling

When the Pilot Program began in late 2005, the four Pilot communities, FHWA, Rails-to-Trails Conservancy, Marin County Bicycle Coalition, and other participants in the program convened a meeting to discuss the evaluation of the Pilot Program in the context of meeting the legislative requirements for reporting to Congress. The group identified the benefits of data collection and evaluation for information sharing and decision making in each community. In addition, the group identified how these evaluation outcome products could contribute to nonmotorized transportation planning, particularly in light of the diverse set of communities participating in the program.

Although the legislation did not provide funding for data collection and evaluation, the Pilot communities realized the need and agreed to work together to collect data and analyze results. The Pilot communities, as a group, set aside a portion of the program funds to effectively meet reporting requirements for this program. FHWA, using its own research funds, also provided support for data collection and evaluation throughout the duration of the Pilot. The US Department of Transportation's Volpe National Transportation Systems Center (Volpe Center) was tasked with coordinating the evaluation process across the Pilots and helped to facilitate partnerships with other partners such as the Centers for Disease Control and Prevention (CDC).

The evaluation framework created by the Pilots focused on legislatively defined goals. Section 1807 of SAFTEA-LU identified eight key program outcomes to be tracked. These were:

1. Frequency of bicycling and walking
2. Public transportation usage
3. Motor vehicle usage
4. Congestion
5. Connectivity to community activity centers
6. Energy usage
7. Environment
8. Health

The evaluation approach was designed to measure these eight large elements. The approach also sought to establish community-wide and project-specific measurements to gauge overall shifts in transportation behavior and impacts tied directly to specific projects at the corridor level. The community-wide measurements included annual bicycle and pedestrian counts, annual bicycle and pedestrian intercept surveys, and community-wide "bookend" surveys. The project-level measurement included annual bicycle and pedestrian counts at target sites, annual bicycle and pedestrian intercept surveys at target sites, detailed data collection about land-use demographics around target sites, and before-and-after facility conditions at target sites.

The evaluation process was composed of three interrelated modeling exercises. Initially there were two model efforts identified to capture these outcomes: a University of Minnesota Center for Transportation Studies (UMN-CTS) community-wide bookend survey, and intercept surveys around manual-count locations developed by Alta Planning and Design. Later the Pilot working group, working with the USDOT Volpe Center staff, developed a third model to overcome unforeseen limitations of the first two. Detailed descriptions of the models below provide insight into the significant challenges of active-transportation measurement.

UMN-CTS Community Household Surveys

Early in the program, the Pilot communities and working-group partners contracted with the University of Minnesota's Center for Transportation Studies (UMN-CTS) in collaboration with NuStats to administer bookend surveys in 2006 and 2010. The survey was designed under direction of UMN-CTS to collect travel behavior data to establish baseline information about travel by bicycling and walking for the four communities. As part of the effort, Spokane, Washington, was selected to be a control community.

There were two survey instruments developed for each community: a short mailer and a longer Internet or telephone follow-up survey. For the follow-up survey in 2010, the research team elected to combine these surveys into a streamlined mailer survey prompting all participants to contribute via Internet or telephone. The survey design called for a minimum of 100 responses for each mode (walking, bicycling, transit, driving). This proved challenging to achieve in the communities with lower population and less prevalence of transit service, and resulted in an oversampling to achieve targets for each mode.

In the end, Gotschi et al. conclude that there was an "inability to detect consistent and statistically significant impacts of the evaluated intervention" from the surveys, resulting in "inconclusive" results. The authors point out that "pre/post evaluations that employ probability-based samples are extremely challenging without the availability of routinely collected data, such as regularly conducted household travel surveys and traffic counts."[21]

Despite the lack of conclusive overall results from the before-and-after surveys, analysis of the baseline surveys highlights the extent of nonmotorized use in the communities and potential limitations of traditional mobility-based analyses of nonmotorized trips. While confidence intervals were not reported for these data, the survey found that in 2006 bicycling and walking comprised 20 percent of total person trips by mode in Minneapolis, but accounted for only 3 percent of the daily per-person mileage. This divergence in the overall number of trips and the relatively small overall mileage estimate results from the short distances that pedestrians and bicyclists generally travel for each trip. The distance for each utilitarian, non-recreation active-transportation trip was estimated and then the resulting estimates were added together to create an overall avoided-driving figure. Combined bicycling and walking was estimated to account for a daily per-person vehicle-mileage reduction of less than a mile (.816) or just over one half a vehicle trip avoided daily.

While the baseline survey showed relatively low vehicle-mileage reductions, the survey simultaneously showed that bicycling and walking already account for one out of five trips for Minneapolis residents. If the policy goal is to facilitate access to destinations rather than pure mobility of residents, it is reasonable to suggest that bicycling and walking already carry a significant part of the transportation load for Minneapolis. Framing the policy success question around access rather than mobility highlights the extent of trips completed by active transportation (20 percent) rather than the pure mileage of active-transportation trips replacing car trips (3 percent). While it is clear that the current congressional policy landscape requires estimation of the mobility-centered benchmarks, increased attention to access indicators is vital for understanding the potential for active transportation to solidify less oil-dependent communities.[22]

Alta Model

Early in the program, the working group became concerned that the community-level survey analysis of UMN-CTS would potentially be too broad to capture the more micro-level changes in active-transportation behavior that they hoped would occur around new facilities. In response, Alta Planning and Design was contracted by the Pilots to develop a community-wide estimate for mode share change and avoided vehicle-miles traveled (VMT) across the Pilots, based on location counts of pedestrians and bicyclists. Each of the Pilots worked with Alta to develop a framework for collecting annual counts and conducting user-intercept surveys following the National Pedestrian and Bicycle Documentation Protocol that had been developed by Alta and the Institute of Transportation Engineers (ITE). Under Alta's direction, the communities were advised to designate at least one count location for every 15,000 people.

The Alta Model used the behavior data collected from intercept surveys and local data variable to extrapolate miles of travel by trip purpose for bicycling and walking based on each community's Census ACS data. Additionally, avoided VMT was calculated based upon comparison of bookend results for utilitarian bicycling and walking trips in order to arrive at likely increases in vehicle-miles shifted from driving. Some limitations in the approach included inconsistencies in collecting intercept survey data, the relatively small sample size of the surveys, and the fairly high ACS margin of error for bicycling and walking rates that could exceed any observed changes in mode share. Overall, the effort to conduct intercept surveys to establish these trip purposes was not as successful as hoped. The Working Group continued to search for a more effective measurement framework.

Working Group Model

Realization that the community-wide results from the Alta and UMN-CTS models were not providing the depth of response to the congressional questions set forth for the Pilot program, the Working Group worked collaboratively to brainstorm some additional measures that could better inform the outcomes of the program. The cornerstone of this final model, the Working Group model, was the use of manual counts conducted in each of the communities over the four years of the Pilot implementation.

Before-and-after counts were conducted at locations receiving new infrastructure treatments. Overall, the counts revealed that across the Pilot communities from 2007 to 2010 bicycling and walking were up 49 percent and 22 percent, respectively (Minneapolis 33 percent and 15 percent).

To translate these numbers into more-nuanced data on the Pilot impacts, the Working Group model used the bookend count data to calculate mode share and VMT averted due to nonmotorized travel. The Working Group approach was designed to compensate for the lack of widespread available data on walking and bicycling use and used a number of "work-arounds" to estimate community levels of active transportation. For example, the Working Group Model used the national statistics for mode share from the NHTS for metropolitan areas of various sizes to establish an assumed baseline for the Pilot communities. This approach has the drawback of underestimating the baseline bicycling and walking for both Marin and Minneapolis, both of which enjoy higher than national averages for bicycling and walking, but does provide a consistent framework to apply across the four communities. Using the NHTS-based mode-share calculations, the model then used the actual observed changes in manual count data to estimate a percentage change in nonmotorized mode share. The model controlled for the number of households in each community and assumed that any increases or decreases in nonmotorized shares could be attributed to corresponding decreases or increases in vehicle trips. Changes in mode share were then multiplied by trip distances by mode in order to estimate VMT averted.

Overall, the Working Group Model found that the program resulted in 37.8 million miles of avoided driving and saved 1.67 million gallons of gasoline in Pilot communities between 2007 and 2010. While there are a number of assumptions that must be operationalized due to lack of widespread availability of active transportation data, the model provides a reasonable starting place for estimating active-transportation use.

Evaluating Success: Creating More-Accessible Communities
Each of the three modeling approaches, while providing comparable VMT figures to calculate numerous congressionally mandated outcomes (including energy savings, reduced transportation costs, and congestion relief), tends to highlight the overall distance of travel as the primary outcome. This approach assumes that a mile walked or biked replaces a mile driven in a one-to-one trade-off. While such an approach is clearly an important element of measuring the success of the program, it undervalues the natural trip elasticity that is influenced by active-transportation mode choice.

For example, say a person is inclined to go out for a meal and perhaps do some shopping or catch a movie. If the community is designed for walking or bicycling, the destinations will most likely be neighborhood-based in scale and the entire round trip may be no more than a few miles. However, if the choice is to drive for this trip, the selected destinations will be influenced by factors related to expected traffic congestion and the cost and availability of parking, with low emphasis on neighborhood proximity. With average auto trip lengths of 9.7 miles,[23] it would not be unlikely for the same auto trip to result in a round trip anywhere from 10 to 20 miles to a larger-scale retail/ entertainment district in the region. Indeed, the decision to walk or bicycle results in a primary emphasis being scale of travel and the access to destinations.

This point is similarly demonstrated through the transit-leverage concept outlined by Newman in chapter 12. Newman argues that denser neighborhoods with multiple destinations result in a leverage effect for each transit trip taken. Calculations by Newman and Kenworthy show that each kilometer of a transit trip replaces between 3 and 7 kilometers of vehicle miles.[24]

This same phenomenon is associated with active-transportation trips. Gotschi and Mills examine how the intersection of increased density and the use of walking and bicycling could help facilitate shorter trips.[25] Utilizing research by Ewing et al. discussed previously,[26] Gotschi and Mills estimate that walking and bicycling can decrease overall driving nationally between 1 and 3 percent for shorter trips of under 15 miles by 2050. This would result in between 13 billion and 32 billion miles of avoided driving.

To provide some context to the potential savings associated with active-transportation leveraging, an exploratory examination was conducted to transform the avoided driving associated with the Pilot program into a hypothetical leveraged estimate. Active-transportation leveraging operates by replacing a potentially much longer car trip with a shorter active transportation trip. The current Pilot avoided-driving calculations do not, however, account for the longer length of average car trips. The current estimates assume a one-to-one relationship between walking and bicycling distances and avoided driving.

In this exploratory analysis, estimated walking and bicycling distances from the Pilot analysis were leveraged by multiplying them by a range of likely driving distances from the average US auto trip length of 9.7 miles to half the average trip length (4.86 miles). In other words, we converted the length of bike trips associated with the Pilot into a transformed figure based on a range of average auto trips. The active-transportation leverage factors used here, especially the lower-end estimate, are generally in line with the transit leverage factors established by Newman and Kenworthy.[27]

In the exploratory analysis, avoided driving associated with active-transportation leveraging in all of the Pilot communities (2007–2010) jumps from 37.8 million miles to a high-end estimate (using average auto-trip length) of 367.4 million miles (table 15.2). Gasoline saving also increases from 1.67 million gallons to a high-end estimate of 16.26 million gallons. Low-end estimates of active-transportation leverage benefits (using half the average auto trip length) result in 183.7 million miles averted and 8.13 million gallons of gasoline saved.

The Minneapolis specific estimates from the exploratory analysis represent about half of the overall Pilot total. They range from a high-end leveraging estimate (2007–2010) of 6.2 million gallons to a lower-end estimate of 3.1 million gallons. This

Table 15.2 Pilot community avoided-driving and active-transportation leverage (2007–2010)

Model	Avoided VMT	Total gallons of gasoline saved in 2010
WG estimate	37.8 million miles	1.67 million gallons
Low active-transportation leverage estimate	183.7 million miles	8.13 million gallons
High active-transportation leverage estimate	367.4 million miles	16.26 million gallons

is compared with the unleveraged Working Group estimate of 642,555 gallons saved (table 15.3).

While this exploratory analysis provides a potential range of benefits associated with active-transportation leveraging, more research is needed to clearly pinpoint the specific leverage potential and establish a definitive active-transportation leverage factor. The current exploratory leveraging example provides only a broad potential range of leveraging benefits that highlight the importance of the intersection of active transportation and denser landscapes for managing oil consumption. While this exploratory work was simply designed to show the conceptual potential of active-transportation leveraging rather than to provide definitive measurements of program effectiveness, the bulk of current research on the value of access for managing VMT lends credence to this type of approach. With the inclusion of more-definitive active-transportation leveraging estimates, this type of approach could prove useful in estimating the value of access-based policy approaches.

Culture Change

Culture Change and NTPP: An Alternative Approach to Policy Evaluation

The traditional active-transportation literature focuses on establishing the impact of new active-transportation facilities on usage rates. These technical impact analyses make up the bulk of the active-transportation policy-analysis literature. This technical approach to evaluating the success of the Pilot program was taken in the first half of this chapter. While detailed quantitative analysis of the impact of new active-transportation infrastructure is a vital element for judging sustainable transportation-policy success, analysis of the implementation of the Pilot and the potential administrative changes within agencies and departments tasked with transportation service delivery provide another important avenue for policy-impact analysis rarely examined within the active-transportation literature.

In 2011, Transit for Livable Communities, the administrative agent of the Minneapolis NTPP, tasked the author with producing a qualitative analysis of the Twin

Table 15.3 Minneapolis avoided-driving and active-transportation leverage (2007–2010)

Minneapolis 2007–2010	Avoided VMT	Total gallons of gas saved
WG estimate	14,521,754	642,555 gallons
Low active-transportation leverage estimate	70,572,651	3.13 million gallons
High active-transportation leverage estimate	141,145,302	6.26 million gallons

Cities Pilot program. The resulting document provided an overview of the key opportunities and barriers uncovered during the administration of the Pilot program.[28] This section presents an overview of the key findings of this analysis.

Culture Change: Key Concept for Decreasing Oil Dependence through
Active Transportation
While the full report cited above provides in-depth analysis of the overall BWTC process, the review for this chapter focuses on the mechanisms for overcoming the "bureaucratic inertia" that significantly decreases acceptance of active-transportation practices.[29] Over and over again, respondents pointed to the need for culture change within the agencies tasked with providing transportation for Minneapolis. The bulk of the stakeholders interviewed for the study were dissatisfied with current transportation policy in the Twin Cities. While this is not entirely surprising, given the selective nature of stakeholders interviewed for the study, the consistent use of the culture-change framework to define the transportation policy problem was suggestive of a strong, widely held belief set among the transportation stakeholders. This need for a broad culture change to facilitate a move toward more-sustainable transportation practices is also explored by Newman in chapter 12.

While the culture-change metaphor was consistently used by respondents, defining its meaning and significance for active transportation is more complicated. Culture change within this context could imply anything from altering the everyday practices within the agencies where transportation decisions are made and carried out, to changing the societal expectations about whether walking and bicycling are "normal" methods to choose for reaching destinations. The following section provides clarification based on the experience of the Minneapolis Pilot.

Culture Change within the Department of Public Works
For the last 75 years, the cultural norms and actions within transportation departments have been centered on the most efficient and safe movement of automobiles through the system. This emphasis of transportation professionals on maximizing throughput of automobiles has resulted in a set of professional standards that have been codified in transportation engineering guidelines. The near-total automobile focus of these guidelines has marginalized walking and bicycling, resulting in active-transportation concerns being considered, as former New York City Traffic Commissioner Sam Schwartz argues, "outside of our job descriptions."[30]

Changing these professional norms and standards to include new-design users, bicyclists, and pedestrians poses both a technical challenge of determining the most appropriate new design standards and also a cultural challenge of broadening the perceived mission of transportation departments to include the special needs of pedestrians and bicyclists. Brown notes that the technical challenges in moving organizations to more-sustainable practices often receive the majority of focus while the equally difficult administrative hurdles of organizational change are often understudied.[31] It is only when both these technical and administrative challenges are addressed that more-sustainable solutions can begin to emerge.

One of the key components of moving transportation organizations forward to more sustainable practices is establishing a culture of innovation that rewards creative problem solving in safely integrating the new-design users (bicyclists and pedestrians) with the traditional-design users of the system (automobile drivers). This challenge of moving from a risk-averse culture to a culture of innovation is a significant hurdle addressed in the organizational change literature.[32]

In Minneapolis, the Pilot resulted in an expansion of the use of a wider range of active-transportation facilities and programming. Szczepanski outlines ten key innovative active-transportation treatments that were implemented or expanded during BWTC (table 15.4).[33] These include lane-width reductions, new bicycle facilities such as bicycle boulevards, bicycle-sharing systems, and a focus on enhancing the development potential of land adjacent to bicycle facilities (trail-oriented development).

These innovative treatments either were not in use or were not in widespread use before the Pilot program. The Bike/Walk Twin Cities (BWTC) program was able to significantly expand the use of these treatments, moving them from innovation to more common practice. The process of the transition of these treatments from innovative to common practice helps to define the key steps of culture change necessary

Table 15.4 Innovative active-transportation treatments

Innovative active-transportation treatments	Used prior to BWTC	Currently used in Twin Cities
Bike-sharing system	No	Yes
Colorized and priority bike lanes	No	Yes
Buffered bike lanes	No	Yes
Road diets	No	Yes
Off-street facilities	Yes	Yes
Bike/walk centers and trail-oriented development	Emerging	Yes
Ubiquitous bike parking	No	Yes
Bicycle boulevards	No	Yes
Improved trail crossings	No	Yes
Advisory bike lanes	No	Yes

in order to transform a community toward a more sustainable and less oil-dependent transportation system.

The process of culture change toward the implementation of these new treatments proved to be difficult. A number of regulatory and cultural barriers had to be overcome to get the Department of Public Works to move toward wider use of these treatments. This policy barrier was frequently mentioned in the Pilot interviews as respondents bumped up against this requirement in attempting to establish changes in lane width, signal timing, and traffic-calming practices.

An example that demonstrates the changes necessary to decrease lane-width requirements to install bike lanes helps to define these barriers. Respondents reported that the engineering standards that govern transportation projects in the Minnesota state-aid system make it difficult to decrease lane widths to accommodate new bicycling facilities.[34] Instead of being able to slightly decrease lane widths to accommodate new bicycle lanes, state-aid standards prescribe minimum lane widths of 11 feet. This lane-width requirement makes retrofitting existing streets with new bicycle facilities extremely challenging in built-out, urban environments. For example, one of the contractors hired to work on the program argued that "The state-aid standards in Minnesota are incredibly antiquated" and are "based on suburban or rural standards."[35] This sentiment was echoed by a professional working on BWTC, who argued that the state-aid standards are "so car-oriented that they aren't realistic." The respondent went on to point out that "the bikes are there today and putting a bike lane in makes it more safe and not less safe."

Despite the potential to increase safety for bicyclists and pedestrians through the realignment of existing space to slow traffic through denser, urban areas,[36] the status quo standards remain a steep barrier. Another professional working on the project pointed out that there is little incentive for local departments of public works to take any perceived risks to push for changes in state-aid standards. The respondent argued that there is "no incentive in state-aid process for the Department of Public Works to take any risks, because their number-one asset is their relationships with other government agencies and they don't want to screw up that relationship over bike lanes."

This intersection of the need to maintain cordial professional working relationships and firm policies on lane widths creates a situation where the burden of proof is entirely on proponents of change. The status quo practices are presumed to be safe and effective unless new studies conclusively prove otherwise. This is an example of the "bureaucratic inertia" that stymies innovation and makes auto-oriented practices the default choice for administrators throughout the country.

To overcome these barriers, concerted and sustained efforts are necessary to dislodge energy-intensive, auto-oriented practices and transform the system into a more multi-modal, sustainable system. Placing the administration of BWTC in the hands of Transit for Livable Communities (TLC) provided a seat at the table for a key change agent. While there is insufficient space for a full discussion of the details of the change process here, TLC's three-pronged approach (outlined earlier in the paper) of maximizing the existing roadway for all users, focusing on creating a regional legacy through planning, data collection, and innovation, and building local and regional capacity helped to provide a platform to change the underlying administrative culture of the transportation agencies. This process helped to provide more latitude for internal champions within the department of public works to help extend innovative infrastructure and planning practices to Pilot and non-Pilot projects. While this process was not without its share of tension, the results of the program speak to the opportunity and need for this type of program in other communities around the country.

Culture Change on the Street: Building Acceptance for Pedestrians and Bicyclists
While moving the transportation bureaucracy toward more-innovative practices is a necessary step in active-transportation culture change, working to expand the acceptability of active transportation among the general public is a corollary step that can significantly expand its use. There are two aspects to working to change acceptability of active-transportation among the general public. The first involves expanding the base of active-transportation "customers" through social marketing. In chapter 12, Newman shows how the TravelSmart program can be an effective tool for building wider acceptance of walking, bicycling, and transit use. This type of social marketing program provides an avenue to reach beyond the already committed active-transportation user and expand the base of potential users much more widely in the community. BWTC used a more direct form of social marketing through their Bicycle Ambassador program, which was designed to help citizens better understand how to engage and use the new active-transportation system.

A second part of culture change among the general public is helping to make active transportation "acceptable" among the broader driving population. If average bicyclists and pedestrians feel uncomfortable and unsafe in interacting with drivers, they are unlikely to use the new facilities no matter how much social marketing encourages them to shift their behavior. Conversely, if drivers feel that cyclists are not legitimate users of system, they may react negatively to changes in roadway design and lobby against expanded bicycle facilities. The well-publicized confrontation over the New York City bike-lane expansion shows the potentially intense passions that can be generated in this change process.[37]

BWTC stakeholders noted that there is a feedback loop between culture change in the Departments of Public Works and culture change among the broader driving population. This twofold approach to culture change involves both working with neighborhoods to build new, acceptable infrastructure and creating an expectation within the general public that bicyclists and pedestrians are legitimate transportation users. One of the professionals working on the project summed up the impact of BWTC on altering the culture by pointing out:

> Culture change happens at the individual level in the reactions of drivers to cyclists. In other cities, there is an antagonistic relationship between cyclists and cars, with bikers taking the lane. . . . Now there are enough cyclists on Lake Street that you don't expect not to see bikes and now it's not an antagonistic relationship. You expect to see cyclists. . . . You see the culture change in Minneapolis, but you don't see it in the nearby suburbs where that change hasn't taken place. Drivers are surprised and there is that antagonistic relationship.[38]

The respondent argued that the "great success" of the Pilot was helping to institutionalize this culture-change process in Minneapolis.

This twofold process of culture change within the departments of public works and with the public at large takes time and can result in tension. Avoiding the "active-transportation culture war" outlined in chapter 16 by Fields, Renne, and Mills requires a delicate balance of leadership that encourages and pushes for change. The potential for this type of change is, however, significant. One of the elected officials reported that BWTC was a "good example of how you can create a catalytic change process." The elected official continued by arguing that, "You've got to resource it sufficiently and give them time to implement and give opportunity to make something happen. If you want to make change, you go with an idea and resource it and then you need someone to defend the idea. . . . We're talking about big changes and you don't get that without resources and some time. . . . If you want change, you need to invest in change."[39]

Discussion: The Policy Mechanisms for Limiting Land-Use and Transportation Choices

The section above highlights the challenges and opportunities of creating the cultural changes necessary to expand active-transportation choices. While the agency practices reflect a culture of automotive-centered everyday practices, the larger policy structures of which they are a part also create significant barriers.[40] One of the significant policy barriers in place is the presumption that the status quo practices reflect the default, safe position that can only be changed through studies that definitively establish "scientific proof" of safety and success for the new, countervailing practices.

While research seeking to establish a causal relationship between changes in the built environment and active-transportation usage can help us understand the precise mechanisms of transportation choice, the high standard of causation is often posited as a precondition for policy changes that would increase active-transportation funding and facilities. This burden of proof to justify departure from status quo spending priorities, while appearing to offer a rational response to spending scarce government resources, is based on the assumption that auto-centered, oil-dependent spending priorities are the default position. Levine (2006) provides a strong counterargument to this position.

Instead of sprawl as the default American landscape choice, Levine argues that the auto-dominated landscape is the active product of policy choices that favor this type of development over other less oil-dependent alternatives.[41] The sprawling landscape, in this reading, results not from some preordained customer preference, but instead from market manipulations that favor sprawl over alternatives. An extensive literature has defined how the interlocking, multi-level policies at the federal, state, and local levels have acted to promote a transportation/land-use monoculture of sprawling suburbs with limited transportation choices.[42]

Levine's work provides a clear articulation of how land-use choice is limited by federal, state, and municipal regulations and transportation policies that limit alternative land-use types. He argues that scientific studies evaluating travel behavior within this type of policy landscape often ignore the underlying policy issue of the dearth of mixed-use landscapes. He argues that "improved scientific understandings of the relationship between land use and travel behavior will not resolve the controversy because travel behavior studies are not designed to shed light on the more fundamental question of why there is so little alternative development to begin with."[43]

While Levine's focus centers on the policy barriers to generating more alternative or mixed-use landscapes, analysis from the Nonmotorized Transportation Pilot Program uncovers a second, interlocking policy barrier that hinders the creation of more pedestrian-, transit-, and bicycle-friendly landscapes. The qualitative research conducted for this chapter extends Levine's argument and identifies a set of transportation regulatory and cultural barriers within transportation departments themselves that limit the availability of more innovative walking and bicycling infrastructure treatments.

When coupled together, the planning failure to facilitate alternative land-development choices and the transportation regulation failure to allow more innovative transportation options combine together to decrease the range of safe, convenient, and accessible transportation options. These interlocking barriers act to limit transportation choice and reinforce the oil-dependent, mobility-based transportation system. From a practical level, the analysis also uncovers how change within these departments can be achieved through consistent, sustained engagement that can, over time, alter agency practices and begin to expand transportation choices. This process of culture change from the mobility-based system to an accessibility-centered approach is a necessary step in tapping the potential of active transportation to manage oil dependence.

Conclusion

While expanding active transportation use in Minneapolis has been challenging, the impact on decreasing oil use has been significant. As Kevin Mills highlights in chapter 10, the current active-transportation mode share in Minneapolis of 20 percent helps to decrease overall oil use by approximately 2.5–3 percent in Minneapolis. With the addition of active-transportation leveraging figures, outlined earlier in this chapter, the potential savings could be even greater.

If the current approach of enhancing active-transportation opportunities for short trips already in place in Minneapolis were replicated at a national scale, important mid-range oil-reduction impacts could be expected. This approach is feasible with current technologies. The barrier to achieving these oil-reduction savings is political and cultural, not technical.

Analysis of the Pilot program situates this cultural problem within the transportation agencies tasked with managing and building the public infrastructure. These agencies continue to view transportation "success" though a mobility lens that privileges solutions designed to spur automotive speed and travel distance over access-based solutions that link transportation consumers more seamlessly to destinations. Because of the siloed nature of transportation decision making, impact assessment of the active-transportation policies often focuses almost exclusively on mobility metrics of the program like VMT avoidance while neglecting the harder to measure, but potentially more powerful community co-benefits of improved access, health, and livability that accrue as the system changes. The burden of proof rests entirely on proponents of change while the status quo funding priorities and practices are assumed to be effective and safe.

Despite these barriers, the analysis of the Pilot also shows some early successes in overcoming these political barriers through a culture-change process within both the transportation agencies and the general public. Moving from an expectation that transportation success is defined by more and more miles of travel to an accessibility focus of providing better and closer linkages to jobs, amenities, and commerce takes time and results in tensions as the underlying mobility culture is challenged. Changing from a known pattern, even a pattern that is as unsustainable as the current one, can create uneasiness and fear. For the accessibility paradigm to firmly take hold, the transition through this culture-change process must be effectively managed in order to minimize and mitigate these tensions.

Analysis of the Pilot program experience in Minneapolis highlights some early lessons on how to mitigate those fears and begin the transition toward a less oil-intensive transportation system. The Minneapolis experience highlights how modest, directed investments can significantly expand active-transportation use. If we replicated currently practiced technology in the core of Minneapolis, we could begin to move toward a less oil-dependent future. Additional research is needed to show how this culture-change process can be scaled up and utilized in multiple communities simultaneously to produce a less oil-dependent future.

Notes

1. See, for example: New York City Department of Transportation, *NYC Commuter Cycling Index* (NYDOT report, New York, 2011); Chicago Department of Transportation, *2009 Bike Counts Project* (City of Chicago Department of Transportation [CDOT], Chicago, 2009), www.cityof chicago.org/content/dam/city/depts/cdot/bicycling/CDOT_bicycle_count_study_2009 pdf, accessed February 25, 2012; Bike Walk Twin Cities, *Pedestrian and Bicycle Count Report: Bike Walk Twin Cities 2009* (report prepared by Transit for Livable Communities, Bike Walk Twin Cities Initiative, Minneapolis, MN, 2010), bikewalktwincities.org/sites/default/files /BWTC2009CountReport.pdf; Portland Bureau of Transportation, *Portland Bicycle Count Report 2010* (report, Portland Bureau of Transportation (PBOT), Portland, OR, 2010).

2. Susan Handy, "Critical Assessment of the Literature on the Relationships Among Transportation, Land Use, and Physical Activity" (report prepared for the Committee on Physical Activity, Health, Transportation, and Land Use, Transportation Research Board, Department of Environmental Science and Policy, Washington, DC, 2005); Lawrence Frank and Sarah Kavage, "A National Plan for Physical Activity: The Enabling Role of the Built Environment," *Journal of Physical Activity and Health* 6, Suppl. 2 (2009): S186–S195.

3. Jonathan Levine, *Zoned Out: Regulations, Markets and Choices in Transportation and Metropolitan Land Use* (report, Resources for the Future, Surface Transportation Policy Partnership, Transit for Livable Communities, Washington, DC, 2006); Minnesota Center for

Environmental Advocacy, *Planning to Succeed? An Assessment of Transportation and Land-Use Decision-Making in the Twin Cities Region* (Washington, DC: Surface Transportation Policy Partnership, 2011).

4. *Planning to Succeed?* 3.

5. Kevin Krizek, Gary Barnes, and Ryan Wilson, *Nonmotorized Transportation Pilot Program Evaluation Study Final Report* (report, Hubert H. Humphrey Institute of Public Affairs, University of Minnesota, Minneapolis, MN, 2007); Kevin Krizek, Ann Forsyth, and Susan Handy, "Explaining Changes in Walking and Bicycling Behavior: Challenges for Transportation Research," *Environment and Planning B: Planning and Design 2009* 36 (2009): 725–40; Thomas Gotschi, Kevin J. Krizek, Laurie McGinnis, Jan Lucke, and Joe Barbeau, *Nonmotorized Transportation Pilot Program Evaluation Study, Phase 2* (report no. CTS 11–13, Center for Transportation Studies, University of Minnesota, Minneapolis, MN, 2011).

6. Federal Highway Administration, *Final Report to the U.S. Congress on the Nonmotorized Transportation Pilot Program SAFTEA-LU 1807* (FHWA report, US Department of Transportation, Washington, DC, 2012).

7. Paul McMorrow, "A Frugal Answer to Zoning Pitfalls, Needlessly Slashed," *Boston Globe*, November 29, 2011.

8. James F. Sallis, Robert B. Cervero, William Ascher, Karla A. Henderson, M. Katherine Kraft, and Jacqueline Kerr, "An Ecological Approach to Creating Active Living Communities," *Annual Review Public Health* 27 (2006): 297–322; Frank and Kavage, "A National Plan for Physical Activity."

9. Ewing Reid and Robert Cervero, "Travel and the Urban Form: A Synthesis," *Journal of Transportation Research Board* 1780 (2001): 87–114; Marlon Boarnet and R. Crane, "The Influence of Land Use on Travel Behavior: Specification and Estimation Strategies," *Transportation Research Part A* 35 (2001): 823–45; Handy, "Critical Assessment"; Levine, *Zoned Out*.

10. Handy, "Critical Assessment"; Frank and Kavage, "A National Plan for Physical Activity."

11. John Pucher, Ralph Buehler, and Mark Seinen, "Bicycling Renaissance in North America? An Update and Re-appraisal of Cycling Trends and Policies," *Transportation Research Part A* 45 (2011): 451–75; Thomas Gotschi, "Costs and Benefits of Bicycling Investments in Portland, Oregon," *Journal of Physical Activity and Health* 8, Suppl. 1 (2011): S49–S58; Kevin Krizek, Ann Forsyth, and Susan Handy, "Explaining Changes in Walking and Bicycling Behavior: Challenges for Transportation Research," *Environment and Planning B: Planning and Design 2009* 36 (2009): 725–40.

12. Robert Cervero and K. Kockelman, "Travel Demand and the 3 Ds: Density, Diversity, and Design," *Transportation Research Part 3* (1997): 199–219.

13. Frank and Kavage, "A National Plan for Physical Activity."

14. Chuck Kooshian and Steve Winkelman, *Growing Wealthier: Smart Growth, Climate Change and Prosperity* (Washington, DC: Center for Clean Air Policy, 2011).

15. Ibid., 19–20.

16. R. Ewing, K. Bartholomew, S. Winkelman, J. Walters, and D. Chen, *Growing Cooler: The*

Evidence on Urban Development and Climate Change (Washington, DC: Urban Land Institute, 2008).

17. Christopher Leinberger, *The Option of Urbanism: Investing in a New American Dream* (Washington, DC: Island Press, 2007).

18. Arthur C. Nelson, "Leadership in a New Era," *Journal of the American Planning Association* 72, no. 4 (2006): 393–409.

19. Sam Schwartz, "A Transportation Engineer's Lament," *Engineering News Record*, December 7, 2011, enr.construction.com/opinions/blogs/schwartz.asp?plckController=Blog&plckBlogPa ge=BlogViewPost&newspaperUserId=00770f29–477f–46d4-bbc3–436b72d800d9&plckPostId= Blog%3a00770f29–477f–46d4-bbc3–436b72d800d9Post%3a85b46226–3500–4e3a–9f9b–8d7638b1 a37d&plckScript=blogScript&plckElementId=blogDest.

20. The Metropolitan Council is the regional planning authority for the Twin Cities area. While it limited planning funding for active transportation, federal funding could still be used for active-transportation project construction.

21. Gotschi et al., *Nonmotorized Transportation Pilot Program*.

22. In a separate study, UMN-CTS (Johnson 2010) present a detailed analysis of access in Minneapolis. This study shows the value of denser communities that bring destinations closer to residents via active transportation, transit, and auto travel; see: Curtis Johnson, "Measuring What Matters: Access to Destinations Research Summary No. 2" (report CTS 10–11, Center for Transportation Studies University of Minnesota, Minneapolis, MN, 2010).

23. Stacy C. Davis, Susan W. Diegel, and Robert G. Boundy, *Transportation Energy Data Book* (publication of the US Department of Energy under Contract No. DE-AC05-00OR22725, Washington, DC, 2011).

24. Peter Newman and Jeff Kenworthy, *Sustainability and Cities: Overcoming Automobile De-pendence* (Washington, DC: Island Press, 1999).

25. T. Gotschi, and K. Mills, *Active Transportation for America—The Case for Increased Federal Investment in Bicycling and Walking* (publication of the Rails-to-Trails Conservancy, Washington, DC, 2008).

26. Ewing et al., *Growing Cooler*.

27. Newman and Kenworthy, *Sustainability and Cities*.

28. Billy Fields, *Building Capacity for Change: Qualitative Impact Assessment of the Twin Cities Nonmotorized Transportation Pilot Program* (unpublished technical report, Transit for Livable Communities, 2011).

29. McMorrow, "A Frugal Answer."

30. Schwartz, "A Transportation Engineer's Lament."

31. Rebekah Brown, "Impediments to Integrated Urban Stormwater Management: The Need for Institutional Reform," *Environmental Management* 36, no. 3 (2005): 455–68.

32. Robert B. Denhardt, Janet V. Denhardt, and Maria P. Aristigueta, *Managing Human Be-havior in Public and Nonprofit Organizations*, 2nd ed. (Los Angeles: SAGE Publications, 2009).

33. Carolyn Szczepanski, "From Minneapolis: Ten Street Design Solutions to Transform

Your City," *Streetsblog Monday*, August 22, 2011, dc.streetsblog.org/2011/08/22/from-minneap olis-ten-street-design-solutions-to-transform-your-city/, accessed February 23, 2012.

34. The State-aid system is a network of state controlled and funded roads. While the vast majority of these roads are in rural areas, a portion of the network extends into urbanized areas. The design standards that are used in rural areas are generally applied in the urbanized areas as well.

35. Quotation drawn from the author's unpublished research.

36. Michelle Ernst, *Dangerous by Design 2011: Solving the Epidemic of Preventable Pedestrian Deaths* (Washington, DC: Transportation for America, 2011).

37. Michael M. Grynbaum, "Lawsuit Seeks to Erase Bike Lane in New York City," *New York Times*, March 8, 2011, A1.

38. Quotation drawn from the author's unpublished research.

39. Ibid.

40. D. Banister, J. Pucher, and M. Lee-Gosselin, "Making Sustainable Transport Politically and Publicly Acceptable," in P. Rietveld and R. Stough, eds., *Institutions and Sustainable Transport: Regulatory Reform in Advanced Economies* (Cheltenham, UK: Edward Elgar Publishing, 2007).

41. Levine, *Zoned Out*.

42. Peter Dreier, John Mollenkopf, and Todd Swanstrom, *Place Matters: Metropolitics for the Twenty-First Century* (Lawrence, KS: University Press of Kansas, 2004).

43. Levine, *Zoned Out*, 23–24.

From Potential to Practice

16

Building a National Policy Framework for Transportation Oil Reduction

BILLY FIELDS, JOHN L. RENNE, AND KEVIN MILLS

The underlying argument of this book is that we currently have the technical capacity to significantly decrease transportation oil consumption by creating a multimodal transportation system. The broadly ranging set of chapters has laid out the significant negative impacts of the current system, the potential of various modes including passenger and freight systems, the need for better connecting transportation with land-use policy, and the potential economic, social, and environmental benefits of moving toward a less auto- and oil-dependent future.

While the technical capacity to address the transportation oil-dependence problem currently exists, the political will to change energy-intensive transportation practices is lacking. Congressional battle lines over transportation policy have hardened in recent years. Recent struggles over reauthorization of the transportation bill point to the emergence of transportation as the latest stage in the seemingly never-ending political culture war.[1] Pitting rural versus urban, wilderness versus growth, driver versus cyclist, and, of course, Democrat versus Republican, the new battle lines in the transportation culture war separate groups in ways designed to suffocate rational debate. Finding a path forward beyond the political camps of the new transportation culture wars to a less oil-dependent society is a vital task that is much more political than technical.

This chapter addresses the key policy barriers to decreasing oil dependence by summarizing the extensive arguments made throughout the book showing that changes to a less oil-dependent system are not only possible, but also advantageous. Instead of being confined by the politics that are driving our nation apart, the policy changes outlined in this book provide an avenue forward that will strengthen quality of life and minimize oil dependence. However, following this avenue will require partnership between both sides of the aisle.

The Arrival of the Transportation Culture War

After ten extensions of the previous transportation bill, Congress passed a new transportation bill in June 2012. The bill, Moving Ahead for Progress in the Twenty-First Century (MAP–21), takes a step backward by reverting to an era when federal policy focused intently on building new roads while neglecting other transportation options such as transit, walking, and bicycling. Aside from maintenance of transit funding at near status quo levels,[2] the bill radically departs from 20 years of reform toward more balanced transportation options by increasing the federal share of road-building costs to 95 percent from 80 percent (but not doing the same for transit) and rejecting reforms that encourage focus on repair of crumbling roads and bridges. MAP–21 also significantly decreases bicycle and pedestrian funding from the already-small percentage allocated to these modes in the predecessor bill, SAFETEA-LU, and simultaneously creates numerous ways for states to avoid spending those resources on walking and bicycling.

Several key episodes during the debate about MAP–21 highlight the political failure of Washington to craft an effective framework for addressing transportation oil dependence and, unfortunately, point to an extension of the culture wars into the transportation arena. The first element of the hardening divide in transportation policy came as representatives from the House Transportation and Infrastructure Committee gathered to craft their version of a transportation bill in November 2011. In previous reauthorizations of the transportation bill, a bipartisan consensus on maintaining funding for transportation was a given. In this new era of constrained finances and hyper-partisanship, the fault lines between Republicans and Democrats for mapping out a national vision for transportation were much more visible.

In addressing shortfalls in the Transportation Trust Fund,[3] House Republicans called for "a massive expansion of offshore oil and gas leasing" designed to "funnel energy-development revenues into infrastructure spending."[4] In discussing the House bill, known as the American Energy & Infrastructure Jobs Act, House Leader John Boehner (R-OH) argued on his official blog posting that the Act would build on the "natural link between energy production and infrastructure energy production" by lifting a ban on new offshore drilling, and opening up US oil shale resources and a portion of the Arctic National Wildlife Refuge in Alaska for energy production. Drawing on the concept of user payments to fund transportation, the bill would also "remove federal requirements that currently force states to spend highway money on non-highway activities."[5] This translates to elimination of transit and active-transportation funding from the transportation bill.

The House Republican plan essentially aimed to tie oil production to oil use. This policy framework would create, as Snyder argued, "a horrific feedback loop" in which the US drills "for oil to pay for infrastructure to drive more cars to burn more oil—it's a recipe to entrench oil dependence in transportation policy in a whole new way."[6]

The effort to expand the "natural link" between oil production and infrastructure policy set the stage for a new front in the culture wars. Instead of the perpetual "environment versus growth" war over expanded oil production in Alaska's Arctic National Wildlife Refuge, the political purpose of the Act was to sharpen the wedge issue by tying the paving of streets in our communities to oil extraction in wildlife areas. To further sharpen the wedge and appeal to cost-conscious conservatives, the Act was to be minimally funded by excluding "non-highway spending" from the list of eligible activities. The Act seemed to be crafted for maximal wedge effect by linking highway funding in our communities to exclusion of funding for "others" (pedestrians, cyclists, and transit riders), and the bill expanded gas exploration in distant wildlife areas "somewhere else."

The House bill was never brought to the floor due to opposition from both the Left and the Right. Criticism of the bill was strong. US Department of Transportation Secretary Ray LaHood, a former Republican congressman who served Illinois' 18th District for 14 years, stated, "This is the most partisan transportation bill that I have ever seen." He also noted that the proposed bill at the time was "the worst transportation bill I've ever seen during 35 years of public service."[7]

While the House bill should have died with its failure in the House, House conferees used it as their script in conference negotiations on MAP–21. In a conference committee of the House and the Senate, various details of competing proposals were hashed out, resulting in a 27-month transportation bill completed in June 2012. House members succeeded in reintroducing a number of major concepts from their bill into the final version of MAP–21. Among these was the House culture-war concept of pitting transportation modes against one another. House Republicans focused on excluding walking and bicycling funding and insisted that funding for active transportation would be a "deal breaker" in the negotiated process.[8] Walking and bicycling funding was decreased by 34 percent in the final bill,[9] and was made subject to further reductions through state DOT opt-outs as well as competition from the addition of expensive new eligible-funding categories.

The framework for making walking and bicycling political fodder for the culture wars was built through a concerted effort to define all non-highway spending as "wasteful." From this perspective, "wasteful" is defined as anything that is not in the core of a narrowly defined federal transportation mission or, in other words, anything that is not more auto-oriented roads and highways. Bicycle/pedestrian and transit programs have been consistently tagged as "peripheral concerns" that should not be funded by the federal government.[10]

While the chapters in this book build a strong case that transit, rail, and active-transportation investments can make a significant contribution to decreasing oil dependence, the political dialogue in Washington defines away these concerns as "peripheral" and centers transportation policy squarely in a 1950s worldview of limitless resources, urban expansion, and environmentally cost-free decision making. For a significant and vocal group in Congress, decreasing American transportation oil dependence is not seen as a goal of transportation policy.

While the transportation culture wars have failed to address the issue of oil dependence in Washington, the need to minimize oil use was made clearer by two reports that emerged during the same time period as the debate about MAP–21. The US Department of Energy calculated the largest single-year increase in greenhouse-gas emissions ever recorded. The Associated Press reported that "the new figures for 2010 mean that levels of greenhouse gases are higher than the worst-case scenario outlined by climate experts just four years ago."[11] Meanwhile, a report from the International Energy Agency projected the "end of cheap oil" and the simultaneous pressure to address climate change to avoid catastrophic "temperature increase(s) of 6°C or more."[12] The chief economist at the International Energy Agency, Fatih Birol, argued that we have less than five years left to change policy direction on energy use. He argued that "if we don't change direction now on how we use energy, we will end up beyond what scientists tell us is the minimum [for safety]. The door will be closed forever."[13]

While time is growing short for avoiding the most serious impacts of climate change, the International Energy Agency points to "the critical role" of energy policy to alter this potential future. The underlying message is that policy matters and that, though time is growing short, it is still possible to steer a course that will avoid the worst environmental and economic shocks of overdependence on fossil fuel.

This all leads to a moment of decision for United States policy makers in terms of the overall direction of energy policy generally and transportation policy specifically. Is the United States ready to begin to move away from oil dependence and toward a more flexible, resilient transportation system, or will US policy makers continue to double down on oil and link future US economic competitiveness to a shrinking and volatile commodity, with huge negative environmental consequences? This is the choice America faces in the coming years, as the successor to MAP–21 will be debated in 2014.

Anatomy of a Policy Intervention: From Efficient, Effective, and Equitable to Politically Feasible

While the path to addressing this policy challenge is complex and contentious, breaking the issue down into its fundamental components can be a useful starting point for outlining potential avenues forward. Basic policy analysis tells us that government intervention to solve a policy problem is often premised on a failure of the free market system to correct a perceived problem and on the capacity of the government to effectively "solve" this problem in an efficient, effective, and equitable manner.[14] This basic framework provides a useful yardstick for judging the potential of alternative responses to address a policy problem.

Taken together, the chapters in this book have created a compelling case for a dramatic course correction in transportation-related energy policy by addressing each one of these issues. The arguments made throughout this book outline the severity of the oil-dependence problem, the failure of the private sector to self-correct the problem, and the potential for government intervention to produce more efficient, effective, and equitable policy responses. The well-documented evidence laid out in the book is outlined below.

Policy Hurdle 1: Significance of the Problem

In a nutshell, the overall policy problem addressed in this book is that the transportation system's dramatic overreliance on oil creates numerous and extensive externalities that threaten long-term environmental and economic stability. Chapters throughout the book have highlighted the dramatic negative consequences of the current system in terms of environmental degradation and potential economic shocks of overreliance on oil.

With 70 percent of the world's oil production consumed in the United States and over one-third of CO_2 emissions coming from the transportation sector, transportation energy use represents a serious driver of significant environmental impacts. In the introduction, we lay out the stark image of the *Deepwater Horizon* disaster and the connection to transportation oil use. The search for oil in the deepest parts of the ocean through the practice of extreme drilling (described in chap. 5 by Lovaas and Potter) is being pushed by the energy demands of the transportation sector.

Once this oil is burned to power vehicles, it produces significant quantities of greenhouse gases. In chapter 1, Gordon and Burwell lay out in detail how the transportation sector contributes to the climate-change problem. They point out that the United States contributes 40 percent of all transportation-sector greenhouse-gas emissions worldwide. They further point out that the transportation sector "is projected to have the greatest influence on the planet's climate at least through 2050." Addressing transportation oil use thus becomes a central early-mitigation strategy in addressing climate change.

In addition to environmental impacts, the economic, social, and human health impacts of transportation oil reliance all pose serious policy challenges. In chapter 2, Sipe and Dodson systematically map what they call "oil vulnerability" in US cities. They discover, not surprisingly, that more sprawling regions are significantly more vulnerable to oil price fluctuations. The fluctuations put increased financial pressure on suburban homeowners as transportation costs increase.

In chapter 3, Litman extends this discussion and summarizes a broad set of externalities associated with transportation oil use. He points out that large direct and indirect subsidies are artificially lowering today's oil prices. He calculates that if the myriad of external costs associated with oil consumption were factored in, the price of a gallon of gas would increase by between $.63 and $1.08. It should be noted that Litman's analysis only factors in external costs of petroleum production, excluding the potentially significant environmental costs of pollution and the human health impacts of traffic crashes.

Taken together, transportation-related oil use creates significant environmental, social, and economic impacts.

Policy Hurdle 2: Entrenched Problems That Will Not Self-Correct
While it is important to show that a problem is significant enough to require intervention, it is equally important to show that the problem will not self-correct over time. At its core, the problems addressed in this book stem from a failure of the private market to develop the ability to transition to a non-oil system and the simultaneous government interventions in the transportation sector that accentuate this market failure.

In chapter 1, Gordon and Burwell argue that it will take no less than a paradigm shift in the way that we view transportation policy to begin a course correction that will address these twin problems. They argue that "the market alone cannot accelerate change in this sector, which is dominated by automobiles as well as the oil companies, institutions, land uses, and lifestyles that support them."

The broad set of institutional actors identified by Gordon and Burwell benefit from extensive subsidies. Litman (chapter 3) describes the extent of the current subsidies that artificially lower gasoline prices. He examines studies that place the extent of these subsidies at between $1.50 and $7.00 per gallon. These subsidies and the failure to include external costs of oil consumption in oil prices simultaneously impact the decisions of private companies about what technologies to invest in and also the decisions that we make as average citizens about where to live and how much to drive.

In addition to the broader economic subsidies embedded in the current system, Noland and Hanson (chapter 4) point to the administrative inertia of the transportation planning sector as a key factor that artificially favors roadway expansion. Noland and Hanson show the impacts of the failure of traffic engineers and policy makers to account for induced demand created by roadway construction. In a groundbreaking, systematic evaluation of research on induced demand, they find "conclusive evidence" that "one cannot reduce congestion through new road projects." Instead, new roads create new demand over time as transportation consumers adjust their behavior based on the increased supply of roadways. These findings call into question the basic premise of road engineers and policy makers that call for increased spending on new capacity as a congestion-mitigation policy.

The results of the Noland and Hanson chapter transform the policy issue from whether to invest in new roadways in order to decrease congestion to "whether development should be dispersed or more concentrated and amenable to nonmotorized modes of travel." In other words, transportation-policy interventions can no longer be justified on the grounds of congestion mitigation, but instead must be resolved based on a vision about how we will shape our communities. The choice is whether we should invest in more oil-intensive or less oil-dependent transportation and community forms. This is the fundamental transportation-policy choice for our times.

While the broad policy choice is clear, shifting agency practices toward less oil-dependent practices is a significant challenge. Fields and Hull (chapter 15) show the difficulties of repositioning the transportation system in a more sustainable manner. They identify the causes of institutional inertia within transportation departments themselves as a factor in maintaining auto-dominated transportation systems and as a barrier to creating more-innovative designs to improve walking, biking, and connections to transit.

Essentially, what we have is an oil-dependent transportation system that has serious negative consequences, but one that is simultaneously vital to powerful economic interests. These interests benefit from extensive government subsidies and agency practices that veil the true costs of the present system. While the negative impacts occasionally bubble up into the popular consciousness through dramatic events such as the *Deepwater Horizon* disaster, they mostly remain background static. When the next inevitable disaster or price upheaval hits, reporters will hit the gas stations of America and find outraged citizens trapped in an increasingly expensive and unsustainable transportation system. From the gas lines of the mid–1970s to the oil disaster in the Gulf in 2010, this has been the story of American transportation policy for the last 40 years. Meanwhile, the problems continue to get worse.

Policy Hurdle 3: Government Intervention Can Be Effective, Efficient, and Equitable

The final, and maybe most significant, policy hurdle addressed is the necessity to show that government intervention can efficiently, effectively, and equitably address the problem. There is widespread acknowledgment that the current oil-dependent transportation system has numerous and serious externalities that negatively impact society as a whole, but there has been to this point no consensus that government intervention can effectively address this massive problem.

The underlying message of *Transport Beyond Oil*, however, is that change to less oil-dependent systems is not only possible, but also economically advantageous. Small changes in multiple sectors can begin to remake the transportation system. Over the course of the next 20 years, these changes can significantly decrease transportation oil consumption and help to effectively, efficiently, and equitably transition the transportation system to a more resilient future.

One of the powerful forces that can be harnessed to help improve quality of life and decrease oil use is the emerging demand for walkable and transit-accessible communities. In chapter 13 Renne discusses the role of transit-oriented developments (TOD) in decreasing oil use. He contends that new TODs not only help to serve pent-up demand but also help to save oil. He examines several different scenarios that show how oil consumption can be significantly decreased by using TODs to meet future housing demand. Oil consumption growth by 2050 in the United States, for example, could be reduced from 40 percent growth over 2010 levels to 29 percent growth if just 30 percent of the future housing growth is built in TODs. This would save 485 million barrels of oil per year. If the nation set an ambitious target of 90 percent of future population growth to reside in TODs, we would be able to reduce oil consumption to 19 percent growth over the 2010 baseline, which would result in 942 million barrels saved per year by 2050.

Renne's findings mirror recent trends that we have seen in what Newman in chapter 12 calls peak car use. Newman argues that we have reached a point where the exponential increases in car use of the post–World War II era have finally stopped and started to decline. Lane, in chapter 6, shows that transit use since 1995 has increased faster than auto travel. Kenworthy's analysis in chapter 14 of transportation data over the last 15 years finds that we are at a tipping point in terms of urban sustainability indicators where cities around the globe are now experiencing increases in density, declines in vehicle-miles traveled, and increases in transit use. The seeds of policy changes planted over the last 20 years are beginning to bear fruit as travel and land-use patterns are beginning to change.

Cities around the globe are beginning to create walkable neighborhoods linked together with transit for longer trips. In the process, they are building the types of communities that are increasingly in demand and are simultaneously decreasing oil use. When these changes are combined with the freight-policy proposals outlined by Drake in chapter 13 and regional rail opportunities (Newman in chapter 14) and/or high-speed rail (Todorovich and Burgess in chapter 10), a full suite of oil-consumption-decreasing policy options can be put in place. These are the type of win-win opportunities that can begin to turn the tide in oil demand.

While recent trends are encouraging in terms of increasing demand for transit-accessible walkable communities, it is vital that affordability of housing options be maintained to ensure equitable access to these new walkable and revitalized neighborhoods. Affordable housing policies and a commitment to quality transit options are central to ensuring that the improvements in quality of life in walkable neighborhoods are not the exclusive purview of the middle class or the rich.

Towards a Less Oil-Dependent Future: Building a Coalition for Change

The ability of government policy to help address a problem in an efficient, effective, and equitable manner is, of course, not the same as the political will to move forward. The concluding section of this chapter addresses how the types of changes outlined in this book can be brought forward.

The opening section of this chapter highlights the growing political divide over transportation policy. As the wedge is sharpened to divide communities, finding the necessary common ground to address such a complex policy problem as oil dependence can often seem like a futile endeavor. The chapters in this book, however, present a different path forward. Small changes in multiple sectors can create the type of changes that can alter the course of the entire system. These changes will be slow and will take a great deal of time to implement. While improved technologies and the specific dimensions of the optimal polices still need to be refined through more research, the outline of a broad policy that can significantly decrease transportation oil use is possible with current technology. The barrier to the widespread adoption of such a policy is political, not technical.

To address political barriers, policies to decrease transportation oil use need to be seen for what they are: proactive, reality-based solutions to difficult problems that can simultaneously improve environmental and neighborhood quality. If policies to decrease transportation oil use are perceived of as threats to quality of life or defined as wasteful government spending, they will become politically toxic. The transportation culture wars are just the beginning wave of this fight over the meaning and future of the shape of our cities and transportation systems.

The good news presented in this book is that, despite deep resistance to change in some areas, communities around the globe are changing to less oil-dependent patterns. They are changing in recognition of environmental constraints, in relation to opportunities that those changes offer in terms of improved quality of life, and in response to overall market forces. Newman in chapter 12 speaks directly to the need for culture change in the way that people understand transportation. He argues that when people begin to make the shift to understanding both the realities of the resource constraints and the potential of improved quality of life, they begin to become advocates for change within their communities. Newman argues that "The politics of change is easier to manage when communities have begun to change themselves."

One of the most effective ways to begin the change is to focus on the qualities of the places that we value most: our neighborhoods. Neighborhoods choked by traffic and exhaust and overwhelmed with speeding vehicles are not the types of places that most people generally want to live. These are not peripheral concerns to neighborhood residents but are instead central concerns that directly impact the quality of life for their families. The challenge is to show how the broader policy structures at the federal and state levels impact those local places and how changing those policies can help improve their quality of life.

Newman begins to show how programs like Safe Routes to School can help to improve neighborhoods by decreasing congestion and building community interactions among parents. The bonds formed in improving neighborhoods through programs like this are the first wave of the culture change necessary to build the political capital to address the larger policy changes necessary to create a far less oil-dependent transportation system. Congressional Representative Earl Blumenaer of Oregon echoes this sentiment when he argues that the importance of Safe Routes to School is "not just as a simple, cost-effective way to improve the safety and health of our children, but as an effective way to engage the public in improving community transportation systems."[15] In this way, building stronger, healthier communities needs to be a central component of any suite of policies designed to decrease oil use.

At the same time, national policies that open up opportunities for these local-level changes need to be given increased attention. Programs such as the Nonmotorized Transportation Pilot Program (discussed in chapter 15 by Fields and Hull), the widely popular TIGER program, New Starts, and the Sustainable Communities Partnership are examples of programs that can begin to grow and build the type of communities and culture change necessary to decrease oil use significantly. Not surprisingly, all of these programs were attacked during the debate on MAP–21 and/or defunded.

This conflict goes to the heart of what the federal role should be in transportation policy. There is a camp that argues that decreasing oil dependence through improved transportation choices is not a federal role. They argue, along the lines outlined in the House Republican American Energy and Infrastructure Jobs Act, that transportation policy should be premised on expanded exploitation of energy reserves and increased highway capacity. Essentially, they are arguing for doubling down on a vision of endless roads and limitless resources. They seem to argue that the only way to solve national transportation problems is by more drilling for oil and more building of roads further into the countryside to alleviate the congestion that their roads inevitably induce.

The problems of transportation oil dependence will not, however, go away on their own by tightening the ideological straightjacket of more drilling and more highway expansion. *Transport Beyond Oil* makes a substantially different case. The problem of transportation oil over-dependence is a national problem that requires national solutions. Implementing these solutions, however, requires not only enhancing regional, interconnected rail networks for long-distance freight and passenger travel, but also focusing on improving access within our metropolitan areas with community-focused initiatives that increase more local transportation choices. If we are to change course, we need to systematically alter our public-policy choices to invest in the types of communities that are in demand in the twenty-first century.

There is a clear federal role in building a twenty-first-century infrastructure system that decreases oil dependence, strengthens economic development opportunities, and ultimately improves community quality of life. *Transport Beyond Oil* begins to lay out this template. It will take step-by-step efforts at multiple levels to begin to push this vision forward.

Notes

1. The culture-war concept basically refers to the emergence of social issues as key political issues. Kahan et al. (2007) argue that the underlying premise of the culture war is really about whose "facts" policies will be based upon. This idea—that there is a "culture war of facts" (p. 2)—is particularly helpful in understanding transportation policy. This book presents a whole set of data to build a case about the need to decrease oil dependence. There is a strongly opposing position that argues that oil dependence is not actually a problem. This argument is built on an entirely different set of "facts" and values that privileges natural resource use over

conservation and growth over long-term sustainability. The transportation culture-war concept discussed here argues that these value issues are becoming central to policy discussions in transportation policy; see: D. M. Kahan, D. Braman, P. Slovic, J. Gastil, and G. L. Cohen, "The Second National Risk and Culture Study: Making Sense of—and Making Progress in— the American Culture War of Fact" (Public Law Working Paper No. 154, Yale Law School, Yale University, New Haven, CT, 2007), papers.ssrn.com/sol3/papers.cfm?abstract_id=1017189.

2. Yonah Freemark, "Congress Passes Major Transportation Bill, Preserving the Status Quo," *The Transport Politic*, 2012, www.thetransportpolitic.com/2012/07/01/congress-passes -major-transportation-bill-preserving-the-status-quo/, accessed July 2, 2012.

3. The issues behind the solvency of the Transportation Trust Fund are covered in chapter 1.

4. Ben German, "GOP Calls for Drilling in Arctic Wildlife Refuge to Pay for Infrastructure," *The Hill*, November 11, 2011.

5. John Boehner, Boehner Blog, American Energy & Infrastructure Jobs Act (H.R. 7), posted by Congressman Boehner's Press Office on November 17, 2011.

6. Tanya Snyder, "Coming Soon: Super-Partisan 'Oil-for-Infrastructure' Transpo Bill," *DC Streetsblog*, November 4, 2011, dc.streetsblog.org/2011/11/04/coming-soon-super-partisan-oil -for-infrastructure-transpo-bill, accessed July 19, 2012.

7. Burgess Everett, "GOP Highway Spending Bill 'The Worst,' Ray LaHood Says," *Politico*, 2012, www.politico.com/news/stories/0212/72369.html#ixzz1ltpJIpnZ, accessed July 15, 2012.

8. Tanya Snyder, "A New Bill Passes, But America's Transpo Policy Stays Stuck in 20th Century," *DC Streetsblog*, June 29, 2012, dc.streetsblog.org/2012/06/29/a-new-bill-passes-but -americas-transpo-policy-stays-stuck-in–20th-century/#more–127085, accessed July 2, 2012.

9. David Burwell, "Buck Up, Reformers: Despite the Hard Knocks, This Bill Is a Step Forward," *DC Streetsblog*, July 5, 2012, dc.streetsblog.org/2012/07/05/buck-up-reformers-de spite-the-hard-knocks-this-bill-is-a-step-forward/#more–127189, accessed July 5, 2012.

10. Robert W. Poole Jr. and Adrian T. Moore, "Restoring in the Highway Trust Fund" (Policy Study 386, Reason Foundation, August 2010), reason.org/files/restoring_highway_trust _fund.pdf, accessed July 2, 2012.

11. Associated Press, "Biggest-Ever Jump Seen in Global Warming Gases," *Los Angeles Times*, November 4, 2011.

12. International Energy Agency, *World Energy Outlook 2011* (report issued by OECD/International Energy Agency, Paris, 2011).

13. Reported by Fiona Harvey, "World Headed for Irreversible Climate Change in Five Years, IEA warns," *The Guardian*, November 9, 2011.

14. M. E. Kraft and S. R. Furlong, *Public Policy: Politics, Analysis, and Alternatives*, 4th ed. (Washington, DC: CQ Press, 2013).

15. Earl Blumenauer, "Consequences of the Worst Bill Ever" (official blog, February 16, 2012), blumenauer.house.gov/index.php?option=com_content&view=article&id=1987:conse quences-of-the-worst-bill-ever&catid=58:op-ed-columns-use-this, accessed July 6, 2012.

Contributors

Billy Fields, PhD, is an assistant professor of political science at Texas State University. His research focuses on understanding the key elements of resilient communities. He has examined resiliency from transportation, urban planning, and hazard mitigation perspectives through positions as the director of research at Rails-to-Trails Conservancy in Washington, DC, and most recently as the director of the Center of Urban and Public Affairs at the University of New Orleans. Recent publications from Fields have appeared in the *Journal of Public Health Policy*, the *Journal of Urban Design*, and *Cityscape*. Fields holds a BA from Trinity University and MPA and PhD degrees from the University of New Orleans.

John L. Renne, PhD, AICP, is an associate professor of planning and urban studies at the University of New Orleans (UNO) and director of the Merritt C. Becker Jr. University of New Orleans Transportation Institute. Renne is a member of the American Institute of Certified Planners (AICP). His research focuses on transportation and land-use planning, including livable communities, sustainable transportation, and evacuation planning. Renne is an author and editor of *Transit-Oriented Development: Making It Happen* (Ashgate, 2009). Since moving to New Orleans in August 2005, Renne has shown leadership in the city's recovery. He serves on a number of local and national boards within the transportation and planning fields. Renne has convened several national conferences and workshops on issues related to transit-oriented development as well as livability, sustainable transportation, and evacuation planning. He is a founding host of the National Evacuation Conference and he chairs the City of New Orleans Sustainable Transportation Advisory Committee. Renne is also the founder and managing director of the TOD Group, an investment and development firm specializing in transit-oriented development.

Edward Burgess received his MS in sustainability from Arizona State University, and his BA in chemistry and environmental studies from Princeton University. He specializes in energy and transportation planning and policy, stakeholder negotiation and communication, life-cycle assessment, GIS and spatial analysis, climate science, and urban ecology. Burgess was a NSF IGERT Fellow and conducted research on policy strategies for urban sustainability, with a focus on transportation planning and greenhouse-gas mitigation. He also worked at Regional Plan Association on the America 2050 project to advocate and plan for regional infrastructure around the country, focusing on high-speed rail.

David Burwell is director of the Energy and Climate Program at the Carnegie Endowment. His work at Carnegie focuses on the intersection between energy, transportation, and climate issues, and on policies and practice reforms for reducing global dependence on fossil fuels. Before joining Carnegie he was a principal in the BBG Group, a transportation-consulting firm that addresses climate, energy, and sustainable transportation policy, with a particular focus on how climate and transportation policies can be better coordinated to promote sustainable development and successful communities. During his career he served as cofounder and CEO of the Rails-to-Trails Conservancy and as founding co-chair and president of the Surface Transportation Policy Project, a national transportation-policy reform coalition. A lawyer by training, he also worked for the National Wildlife Federation as director of its Transportation and Infrastructure Program. He has served on the Executive Committee of the National Research Council's Transportation Research Board (1992–1998) and is presently on the board of advisers of the Institute for Transportation and the Environment at the University of California at Davis. He served in the Peace Corps in Senegal, West Africa.

Gilbert E. (Gil) Carmichael, a leading international authority on railroad and intermodal transportation policy, is the founding chairman and serves as a member of the board of directors at the University of Denver's Intermodal Transportation Institute. Carmichael served as the US Department of Transportation Federal Railroad Administrator (FRA) in the administration of President George H. W. Bush from 1989 to 1993, and served as vice chairman of the board of WABTEC Corporation, the leading independent manufacturer of after-market locomotive component parts and the leading independent locomotive remanufacturer in North America. A graduate of Texas A&M University and a former fellow in the Kennedy School of Government at Harvard University, he presents and publishes papers on the transportation industry, promoting the need for a North American and global intermodal freight and passenger system that utilizes the world's rail network.

Jago Dodson is the director of the Urban Research Program at Griffith University. Following a PhD at Melbourne University, where he investigated public housing policy, he has published extensively on a wide range of urban problems ranging across planning, housing, governance, transport, infrastructure, energy use, climate change, and social disadvantage. He has also been a regular media commentator on urban issues in Australia and an advocate for improved policy making in Australian cities. His work with co-author Neil Sipe was widely reported in print and broadcast media during 2007–2010. Dodson has provided research-based advice to local, state, and federal governments.

Alan S. Drake is a consulting engineer who became concerned about both climate change (which he prefers to call climate chaos) and the advent of extreme oil—oil that is ever-more-expensive and difficult to extract in limited quantities. Working *pro bono publico*, he has researched possible mitigation strategies, and their economic consequences, for both issues for over a decade.

Projjal K. Dutta is the Metropolitan Transportation Authority's first-ever Director of Sustainability. He has two primary responsibilities: to reduce the environmental footprint of the MTA, and to verifiably measure the carbon benefits that accrue to the region, due to the MTA. In a carbon-constrained future, this could generate badly needed additional resources. Dutta was instrumental in the measurement and verification of MTA's carbon footprint and its registration with the Climate Registry. He has played a leadership role in the transit industry's effort to quantify its carbon benefits. He has lectured and written extensively on the subject of "carbon avoidance" at Harvard, Yale, and Columbia Universities as well as others. Dutta has more than 20 years' experience in projects ranging in scale from urban to residential, with a particular emphasis on sustainable design. Before joining the MTA, he worked as a sustainable architecture consultant. His graduate thesis, which explored the construction of low-cost housing from waste packaging, was adjudged "Best Thesis" at MIT. His built projects have been featured in publications in the United States and abroad.

Deborah Gordon is a nonresident senior associate in Carnegie's Energy and Climate Program, where her research focuses on climate, energy, and transportation issues, with a special focus on unconventional oil and fossil fuels in the United States and abroad. Since 1996 she has been an author and policy consultant specializing in transportation, energy, and environmental policy for nonprofit, foundation, academic, public, and private sector clients. From 1996 to 2000 she codirected the Transportation and Environment Program at the Yale School of Forestry and Environmental Studies, and from 1989 to 1996 she founded and then directed the Transportation Policy Program at the Union of Concerned Scientists. Additionally, Gordon has worked at the US Department of Energy's Lawrence Berkeley Laboratory (1988–89), where she developed clean-car "feebate" policies under a grant from the US Environmental Protection Agency, and she was a chemical engineer with Chevron (1982–87). Gordon has served on National Academy of Sciences committees and the Transportation Research Board Energy Committee. Her recent book *Two Billion Cars* (co-authored with Daniel Sperling) provides a fact-based case and a roadmap for navigating the biggest global environmental challenge of this century—cars and oil (Oxford University Press, January 2009).

Christopher S. Hanson joined the Voorhees Transportation Center in 2009 as a postdoctoral research associate to work on the Center's Carbon Footprint Project. Through his efforts, this project has developed a methodology for estimating the global warming potential of transportation capital projects for the New Jersey Department of Transportation. Hanson received his PhD in planning and public policy from the Bloustein School for Planning and Public Policy at Rutgers University in 2007. He has also received a MA in geography from Hunter College of the City University of New York in 2001. His research interests include emissions modeling, evaluation and outcomes analysis, and transportation and health policy. He is a past recipient of a cancer-policy graduate fellowship from the New Jersey Commission on Cancer Research, and is a current member of the American Planning Association and the Association of American Geographers.

Tony Hull currently works for Transit for Livable Communities (TLC) in Saint Paul, Minnesota. His responsibilities include planning and evaluation of the Federal Nonmotorized Transportation Pilot Program (SAFETEA-LU 1807), known locally as the Bike Walk Twin Cities program. Prior to coming to TLC, Hull worked as a principal planner at the Mid-Ohio Regional Planning Commission (MORPC) in Columbus, Ohio, and before that he worked with the Central Ohio Transit Authority and the city of Hilliard, Ohio. His background includes over a decade of working as a multimodal transportation planner, with a focus on pedestrian needs, ADA accessibility, Complete Streets, and traffic calming. A graduate of the Ohio State University, Hull has been a member of the Association of Pedestrian and Bicycle Professionals (APBP) since 2004, and currently he serves appointments to the TRB Committee on Pedestrians, ANF10, and the Minneapolis Pedestrian Advisory Committee, representing the Ward 6 neighborhood of Whittier, where he currently resides.

Jeffrey Kenworthy is professor of Sustainable Cities at the Curtin University Sustainability Policy Institute (CUSP) at Curtin University in Perth, Western Australia. He has spent 33 years in the transport and urban planning field and currently teaches courses and supervises postgraduate students in the city policy and urban sustainability fields. Further, he is author and co-author of over 200 books, book chapters, and journal publications in the area of city policy. He has extensive experience in the fields of compact-housing developments, public-transport systems, and sustainable transport policy, and he has worked as a consultant for local, state, and federal governments in Australia, as well as private organizations and the World Bank. For three and a half years he was project director for the Millennium Cities Database for Sustainable Transport for the International Union (Association) of Public Transport in Brussels (UITP). This study includes 100 developed and developing cities in every part of the world and includes comparative data on urban land use, transport, economics, and the environment of cities. Kenworthy received the Australian Centenary Medal from the Australian Prime Minister's Office for service to planning and sustainability in relation to public transport and urban form.

Bradley W. Lane is an assistant professor in the Master of Public Administration Program at the University of Texas at El Paso. His research focuses on travel behavior, policy, and planning issues in urban transportation, and he has taught classes in research methods, urban and regional planning, policy analysis, human geography, and environmental impact. He has conducted and published extensive research on the impact of gasoline prices on public transportation, and on policies, perception, knowledge, and travel behavior related to the electric vehicle. Previously published works also include studies on the renaissance of rail transit in the United States, the spatial variability of changes in travel behavior related to light-rail infrastructure, and on high-speed intercity rail transportation. Lane is a board member of the Transportation Geography Specialty Group of the Association of American Geographers, and also serves as a reviewer for numerous major journals, including the *Proceedings of the National Academy of Science* and the *Transportation Research Board of the National Academies*. Prior to joining the faculty at UTEP in August 2010, Lane earned a PhD in geography from Indiana University in 2010, an MA in geography from Indiana University in 2006, and a BA in political science and public policy from Rice University in 2003.

Jie (Jane) Lin is an associate professor in the department of civil and materials engineering and a research associate professor in the Institute for Environmental Science and Policy at the University of Illinois at Chicago. She is also an affiliated faculty member at the Urban Transport Center (UTC) at UIC, and is a current member of the Transportation and Air Quality Committee (ADC20) of the Transportation Research Board, National Academies. Lin's research interests are in the areas of intelligent transportation systems, sustainable transportation systems, public transit planning and operations, and goods-movement analysis. Lin received her PhD from the University of California at Davis.

Todd Litman is founder and executive director of the Victoria Transport Policy Institute, an independent research organization dedicated to developing innovative solutions to transport problems. His work helps to expand the range of impacts and options considered in transportation decision making, to improve evaluation methods, and to make specialized technical concepts accessible to a larger audience. His research is used worldwide in transport planning and policy analysis. Litman has worked on numerous studies that evaluate transportation costs, benefits, and innovation, and he is active in several professional organizations including the Institute of Transportation Engineers and the Transportation Research Board (TRB, a section of US National Academy of Sciences). He currently chairs the TRB Sustainable Transportation Indicators Subcommittee.

Deron Lovaas is the Natural Resource Defense Council's director of federal transportation policy. His main focus is on policies about transportation and energy, and their effects on public health, the environment, and the economy. He is an expert on a variety of issues, having testified multiple times before Congress on topics including dependence on oil, energy efficiency, fuel economy, transportation infrastructure, roads and bridges, public works, and gasoline taxes, as well as aviation, bus, railroad, bicycle, and pedestrian projects. Prior to his decade at NRDC, Lovaas spent ten years working at several conservation groups including the National Wildlife Federation, Zero Population Growth, and the Sierra Club, where he directed a national campaign to reduce suburban sprawl. He was also an environmental specialist with Maryland's Environment Department from 1993 to 1995. He received a bachelor's degree from the University of Virginia in 1992.

Simon McDonnell is a senior policy analyst for the Office of Policy Research at the City University of New York (CUNY). Previously, he was a research fellow at the Furman Center for Real Estate and Urban Policy at New York University, where his research focus was on land-use, environmental, and transportation policy. He graduated with a BA in economics, an MS in environmental economics and policy, and a PhD in planning and environmental policy from University College, Dublin. McDonnell was also Irish coordinator of a trans-European research project investigating transport sustainability—TranSust.Scan. Before joining the Furman Center, McDonnell spent a year as a visiting assistant professor with the Urban Planning Program and the Institute of Environmental Science and Policy (IESP) in the University of Illinois at Chicago.

Kevin Mills is the vice president for policy and trail development at the Rails-to-Trails Conservancy. He oversees RTC's program agenda, including federal and state legislation and rulemaking, grassroots movement building, program initiatives, and research, and is a national leader in the effort to ensure that trails, biking, and walking remain key elements of America's transportation policy. Prior to joining RTC in spring 2006, Mills spent more than 15 years at Environmental Defense, where he directed programs to reduce the climate and health impacts of automobiles, reduce the use and waste of toxic chemicals, and promote sustainable transportation and communities.

Peter Newman is the professor of sustainability at Curtin University and director of the Curtin University Sustainability Policy Institute. Newman is on the board of Infrastructure Australia, which funds infrastructure for the long-term sustainability of Australian cities, and is a lead author for transport on the IPCC. He has three recent books: *Technologies for Climate-Change Mitigation: Transport* for the UN Environment Program, *Resilient Cities: Responding to Peak Oil and Climate Change*, and *Green Urbanism Down Under* for Island Press. Newman directed the production of Western Australia's sustainability strategy in the Department of the Premier and Cabinet, the first state sustainability strategy in the world, and was a Fulbright Senior Scholar at the University of Virginia, Charlottesville. In Perth, Newman is best known for his work in saving, reviving, and extending the city's rail system. For the 30 years since he attended Stanford University during the first oil crisis, he has been warning cities about preparing for peak oil. Newman's book with Jeffrey Kenworthy, *Sustainability and Cities: Overcoming Automobile Dependence*, was launched in the White House in 1999. From 1976 to 1980 he was a councilor in the city of Fremantle, where he still lives.

Robert B. Noland is the director of the Voorhees Transportation Center (BSPPP) and a professor at the Bloustein School of Planning and Public Policy at Rutgers University. He received his PhD in energy management and environmental policy at the University of Pennsylvania. Prior to joining Rutgers, he was reader in transport and environmental policy at Imperial College, London, and a policy analyst at the US Environmental Protection Agency; he also conducted postdoctoral research in the economics department at the University of California at Irvine. The focus of Noland's research is the impacts of transport planning and policy on environmental outcomes. His research has been cited throughout the world in debates over transport-infrastructure planning and environmental assessment of new infrastructure. Noland is currently the associate editor of the journals *Transportation Research Part D (Transport and Environment)* and the *International Journal of Sustainable Transportation*, and he is chair of the Transportation Research Board Special Task Force on Climate Change and Energy.

Joanne R. Potter is a principal at IFC International, and has more than 15 years' experience in climate change, sustainability, and transportation. Her work has addressed both strategies to reduce greenhouse-gas emissions from transportation and also climate impacts and adaptation analysis. She is currently supporting the US AID program in developing guidance on climate-change impacts analysis and adaptation for infrastructure in the developing world, and the Strategic Environmental Research and Development Program of the US Department of Defense (DoD) in its examination of climate-change impacts on coastal installations. Prior to joining ICF, Potter was a lead and editing author of *The Impacts of Climate Change and Variability on Transportation Systems and Infrastructure: Gulf Coast Study Phase I*, released in March 2008, and was project manager for the US Department of Transportation's (DOT) "Report to Congress on Transportation's Impact on Climate Change and Solutions." She also managed the development and publication of *Moving Cooler: An Analysis of Transportation Strategies for Reducing Greenhouse Gas Emissions* (July 2009), a national multisponsor study assessing the effectiveness of transportation activity strategies to reduce GHG emissions. Potter received a master's degree in city planning from the Massachusetts Institute of Technology and a BA from the University of Massachusetts at Amherst.

Neil Sipe is the deputy director of the Urban Research Program at Griffith University. Since receiving his PhD in urban and regional planning from Florida State University, he has taught in the Griffith University (Brisbane, Australia) School of Environment since 1998 and has served as head of the urban planning program from 2002 to 2006 and 2008 to 2012. He has also served as deputy director of the Urban Research Program from 2006 to 2008 and as acting director from June to December of 2011. Sipe currently serves on the Transportation Research Board Ferry Committee and the National Education Committee of the Planning Institute of Australia and is the editor of *Australian Planner*, a peer-reviewed journal serving Australian planning academics and professionals. He has an extensive teaching record in the field of transport planning and in recent research has proposed methods (with chapter co-author Jago Dodson) for defining and mapping transport exclusion and oil vulnerability. He has a strong track record in empirical research that links issues of spatial access and socioeconomic equity in urban contexts, both in the United States and in Australia.

Petra Todorovitch is a visiting assistant professor at the Graduate Center for Planning at Pratt Institute. She is also the director of America 2050 at the Regional Plan Association, a national urban planning initiative to develop an infrastructure plan and growth strategy for America in the twenty-first century. Todorovich oversees America 2050's research, advocacy, and planning, in partnership with organizations such as the Lincoln Institute of Land Policy and the Rockefeller Foundation. Prior to the launch of America 2050, Todorovich directed the Regional Plan Association's Region's Core program and coordinated the Civic Alliance to Rebuild Downtown New York, a network of organizations that came together shortly after September 11, 2001, to promote the rebuilding of the World Trade Center site and Lower Manhatttan.

Index

Page references followed by "f" refer to text figures; page references followed by "t" refer to text tables

About Island Press

Since 1984, the nonprofit Island Press has been stimulating, shaping, and communicating the ideas that are essential for solving environmental problems worldwide. With more than 800 titles in print and some 40 new releases each year, we are the nation's leading publisher on environmental issues. We identify innovative thinkers and emerging trends in the environmental field. We work with world-renowned experts and authors to develop cross-disciplinary solutions to environmental challenges.

Island Press designs and implements coordinated book publication campaigns in order to communicate our critical messages in print, in person, and online using the latest technologies, programs, and the media. Our goal: to reach targeted audiences—scientists, policymakers, environmental advocates, the media, and concerned citizens—who can and will take action to protect the plants and animals that enrich our world, the ecosystems we need to survive, the water we drink, and the air we breathe.

Island Press gratefully acknowledges the support of its work by the Agua Fund, Inc., The Margaret A. Cargill Foundation, Betsy and Jesse Fink Foundation, The William and Flora Hewlett Foundation, The Kresge Foundation, The Forrest and Frances Lattner Foundation, The Andrew W. Mellon Foundation, The Curtis and Edith Munson Foundation, The Overbrook Foundation, The David and Lucile Packard Foundation, The Summit Foundation, Trust for Architectural Easements, The Winslow Foundation, and other generous donors.

The opinions expressed in this book are those of the author(s) and do not necessarily reflect the views of our donors.